Interpreting the Evolution of Atayal Tribal Textile Patterns

Po Hsun Wang · Jie Li

Interpreting the Evolution of Atayal Tribal Textile Patterns

Tracing the Anthropological Footprints of a Taiwanese Ethnic Minority Group

Springer

Po Hsun Wang
City University of Macau
Taipa, Macao

Jie Li
City University of Macau
Taipa, Macao

ISBN 978-981-96-5450-5 ISBN 978-981-96-5451-2 (eBook)
https://doi.org/10.1007/978-981-96-5451-2

The original submitted manuscript has been translated into English. The translation was done using artificial intelligence. A subsequent revision was performed by the author(s) to further refine the work and to ensure that the translation is appropriate concerning content and scientific correctness. It may, however, read stylistically different from a conventional translation.

This Springer imprint is published by the registered company Springer Nature Singapore Pte Ltd.
The registered company address is: 152 Beach Road, #21-01/04 Gateway East, Singapore 189721, Singapore

Preface I

Ethnic objects result from production and manufacturing; the production method is the most basic cultural phenomenon, and the decorative form on the object's surface, under certain conditions, is a form of art. The expression of traditional costumes of Taiwan's minority ethnic groups implies decorative thinking in the development process of primitive art, which means the technical relationship of craft production and the logical relationship of pattern design. Taking the Atayal tribe as an example, the shape and color of the textile patterns symbolize the cultural characteristics of the ethnic group. Whether it is the structural form or the color combination, to some extent, there is a design thread to follow, which is also the unique clothing style of the ethnic group. Therefore, this study not only needs to clarify the structural form and color combination of the patterns, but another focus is to explore the design thinking of the woven patterns, summarize the various styles of textile patterns, and, even through analysis, construct the possibility of its pattern module system.

The object of this study is the traditional costume of the Atayal tribe, with the "shape and color of Atayal textile patterns" as the research content. The overall structure of the thesis is explained as follows:

Chapter 1: Introduction

The traditional clothing of minority ethnic groups has unique cultural symbols, especially in clothing production. The patterns formed by dyeing and weaving may have become one of the characteristics distributed among the ethnic groups through historical changes or ethnic exchanges. Taiwan's minority ethnic groups, with the broadest geographical distribution of the Atayal tribe, and the patterns of clothing or other textiles are also the most abundant. Pattern shape and color combinations are mainly composed of geometric patterns and hide a systematic organization of patterns. Therefore, through the study of Atayal textiles, in addition to understanding the relationship between the structure, shape, and color of Atayal weaving, on the other hand, it can also integrate the material culture, ethnic art, and textile characteristics of the Atayal tribe, providing references for design and craft.

Chapter 2: Taiwan's Minority Ethnic Groups and the Pan-Atayal Tribe

Reviewing important literature and related theories, understanding the evolution of Taiwan's minority ethnic group classification, and discussing the research results of minority ethnic art characteristics, thereby confirming the research positioning and devising research methods. Explore the material culture of the Pan-Atayal tribe, understand the geographical distribution and ethnic characteristics of the Atayal system from an anthropological perspective, and understand the form characteristics displayed by Atayal textiles. Attempting to start from the visual design standpoint by combining the visual design language with the knowledge categories of decorative patterns and organizational structure through the design thinking of form modules, explore the composition method of textile patterns, and then discover the systematic rules of Atayal weaving patterns.

Chapter 3: Analysis of Traditional Textiles of the Pan-Taiya Ethnic Group

Past field surveys and measurements of Taiwan's minority ethnic groups mainly started from an anthropological perspective, comprehensively recording the society, material, events, etc., of minority ethnic groups. However, the clothing survey only described the appearance of the clothing, so a deeper analysis of textile patterns and colors is needed. Given this, to clarify the complex artistic structure of material culture, it is necessary to combine modern research technology and reconstruct the subtle features of patterns through digital drawing, supplemented by pattern analysis, color comparison, statistical analysis, and other methods, to analyze the clues missed by previous research surveys. In addition, the purpose of finely depicting the textile patterns is not only to explore but also to construct and reproduce the form of Atayal textile patterns, which is one of the most important contributions of this study.

Chapter 4: Characteristics of Pan-Atayal Group Textile Types

Integrating literature descriptions and research samples, the subgroups of the Atayal tribe's textiles are summarized, such as the Hakul, Makanji, Msbtunux, Klesan, Malepa, and Taranan groups of the Tayal-Seqoleq subgroup of the Atayal tribe; the Klesan, Plnawan, Bunsui, Beishi, Nanshi, and Meicyubus groups of the Tseole; the Musha, Pulevao, and Truku groups of the Seediq. For these Atayal subgroups, the patterns and colors on the textiles are mapped and compared, and the materials and weaving techniques are recorded. The Atayal textiles from different collection units reveal the characteristics of the textiles of different subgroups and explore the relationships between the subgroups, providing information for subsequent related research.

Chapter 5: Characteristics of Pan-Atayal Textile Patterns

What patterns are present in the composite stripes that make up Atayal textiles, as revealed by research samples? Understanding the shaping features of pattern types, analyzing the core motifs that make up the patterns, and identifying the basic shapes. Using a method of analysis that connects graphics, we can understand the overall arrangement of the patterns. Regarding the color performance of the composite stripes, the color combinations of Atayal textiles are organized, and the color distribution of subgroup textiles is clarified. This research process can better understand textile patterns' shaping structure and color combination.

Chapter 6: Pan-Atayal Group Textile

Through analyzing specific examples, we explore the formal logic of Atayal textile patterns and the arrangement rules of composite stripes, deducing the relationship between textile patterns and colors and understanding the color composition method of textile patterns. The research findings on Atayal textile patterns, such as comparing the flow relationship between textile patterns and sub-tribal groups, are compiled to construct the geographical distribution of textile patterns, which can be used as corroborative evidence for understanding tribal migration. This research can describe the design's main features and development rules from the diverse Atayal textile patterns.

Chapter 7: Conclusion

The Atayal weaving craft can also be said to be the intangible cultural heritage of Taiwan's minority ethnic groups. The study of the shaping and colors of Atayal textile patterns hopes to draw attention to the unique material culture of minority ethnic groups and use it as a will for generational inheritance. This research intends to establish a combination of design studies and cultural anthropology, reflecting the interdisciplinary nature of design studies. It is hoped that practical design research can be directly applied to authentic design and actual production. The detailed structure of textile patterns depicted in the research process can provide immediate practical value for professionals in the design field while also helping those who understand Atayal culture through traditional patterns to master the design performance of the Atayal style.

This study has completed the research and analysis of the Atayal tribe's textile patterns, hoping to influence researchers interested in the material culture of Taiwan's minority ethnic groups to explore related topics jointly. Many scholars are trying to break free from the geographical constraints of Taiwan's minority ethnic groups and delve into the traditional material culture of the Austronesian ethnic groups. Therefore, future research directions also hope to involve the textile patterns of the Austronesian ethnic groups, thereby comparing the similarities and differences with the textile patterns of Taiwan's minority ethnic groups. In addition, there are still many minority ethnic groups within China whose textiles are rich in decorative solid features, such as the Miao and Dong ethnic groups. These ethnic groups with unique cultural backgrounds are also worth paying attention to, especially in the face of the development of modern society, where traditional, unique cultures may gradually disappear. If they can be recorded through research, this is undoubtedly the field focused on by design humanities and arts.

Macau, China

Po Hsun Wang
Jie Li

Preface II

After about six years, from topic selection, research, and then entering the museum to collect samples, the entire research process was not easy, even spanning different places, and finally completed the study of the material culture of the Atayal tribe, a minority in Taiwan.

Recalling the situation at that time, following the research team, under the leadership of Professors Zeng Qixiong, Hong Minghong, Principal Takahito Shimoki of Kobe University of Art and Technology, and two teachers, Huang Guobin and Cai Yushan, we began to conduct field investigations on minority topics in Taiwan. But at that time, I was confused and had endless doubts. Why do we need to investigate Taiwan's minorities? This topic has many research results; are there still undiscovered ones? With a doubtful heart, the research team, the mood, also went through countless struggles and panics until facing many complicated research topics of Taiwan's minorities; they finally chose Atayal textiles as the research topic and then began to love the textiles of minorities. However, the focus on the material culture research of Taiwan's minorities is also hoped to connect the vision of design studies with material culture research and promote the characteristics of humanities and arts in Taiwan.

The Atayal people each have different daily tasks, with men responsible for hunting, fighting, and farming and women accountable for weaving, cooking, child-rearing, sowing, and harvesting. Weaving is the duty of Atayal women, and the traditional horizontal weaving loom is made by the women's fathers and elders or commissioned by skilled makers. It is worth noting that while men are not supposed to be familiar with weaving techniques, the tools are made by men, suggesting that these cultural objects are produced jointly by both sexes after communication and coordination. The Atayal use traditional clothing as part of their cultural heritage. In cultural transmission, the objects represented by the woven patterns have transcended physical reality, mainly associated with ancestral spirit beliefs (legends), spiritual life, and social power, and are continuously changing. Abstract geometric patterns in woven textiles cannot exist independently but are attached to clothing. In the primitive cultural stage, geometric patterns have important life significance, derived from the ethnic understanding of natural phenomena, and then obtain primary refinement, forming symbolic characteristics. The patterns in Atayal textiles are the result of careful design by women, and the style of the patterns is determined by the form, basic shape, and composition of the composite stripes, with women dominating the overall artistic expression. Compared to other minority ethnic groups in Taiwan, the Paiwan have a more balanced artistic creation, with men specializing in wood carving and house building; women are good at embroidery and sewing, and their tribal environment displays aesthetic experience and creative potential everywhere, seemingly having a well-established division of labor system. Yet, the style of the ethnic group is so harmonious. Perhaps in the future, we can compare the artistic expressions of the Atayal and Paiwan, conduct comparative research on the differences in the interpretation of patterns by different ethnic groups, and even study from a gender perspective how the division of labor between the sexes affects the development of ethnic art.

After completing the Atayal research report and several years of sedimentation and reflection, I am re-examining these research results, compiling and reviewing the graphics and text, and finally sharing them as a publication. Looking back on the entire research process, I sincerely thank Professors Zeng Qixiong, Hong Minghong, Huang Guobin, and Cai Yushan for their careful and patient guidance. The bits and pieces of this research paper should be credited to the assistance of the research team, especially Professor Zeng Qixiong's teachings from topic selection and research to the writing process, leading to reflection on the relationship of design culture research from the research standpoint.

I want to thank Tenri University's Tenri Reference Museum in Japan, the Shung Ye Museum of Formosan Aborigines, the Changhua Fenyuan Showa Hall of Ancient Art, as well as the Taiwan Museum of Prehistory, the Taiwan Museum, the Department of Anthropology at National Taiwan University, the Huludun Cultural Centre and other institutions for providing the necessary research samples and assisting in the application and review of textile research. Also, many friends from Taiwan's minority ethnic groups have helped during field investigations. This research result could not have been achieved without the help of many people, and I would like to express my sincere thanks again.

Zhuhai, China
June 2023

Po Hsun Wang

Acknowledgments I am grateful to the teachers who have guided my research and my wife, who has been my most vital support. This book is also dedicated to my dearly missed father. During this research, I sincerely thank the institutions that have collected traditional costumes of Taiwanese ethnic minorities for their varying degrees of assistance and support.

Competing Interests The authors have no competing interests to declare that are relevant to the content of this manuscript.

Synopsis

Design studies have entered the field of ethnic culture and minority studies from a cross-disciplinary perspective of humanities and social sciences, becoming a topic of interest for many contemporary researchers. Human beings are distinguished by their ability to create tools, transform nature, and use language. In this process of value creation, culture emerges and evolves. Therefore, to understand and learn about a culture, one must understand the characteristics of those who create that culture. Crafts and folk art creation have developed alongside the emergence of culture, becoming important components and expressions of that culture. They are not only expressions of culture but also key windows for understanding and interpreting the culture of an ethnic group. Through these works, one can gain deep insights into the group's values, history, lifestyle, and cultural exchanges, providing rich materials and perspectives for a comprehensive understanding of their culture.

The Atayal people are one of the most representative minority groups in Taiwan, China. Among the indigenous tribes of Taiwan in China, the Atayal are mainly known for their distinctive clothing. Their traditional attire is characterized by vibrant colors and intricate patterns, commonly featuring red, black, and white. Men wear shorts and sleeveless tops, while women wear long skirts and shawls. The designs and motifs on their clothing are aesthetically pleasing and carry symbolic meanings, often representing identity, status, and tribal affiliation. Studying the Atayal people of Taiwan holds significant academic and cultural value. Their folk crafts and traditional attire are masterpieces of artistic creation and crucial carriers of cultural heritage. By examining the history, language, social structure, and cultural arts of the Atayal, one can gain a deeper understanding of Taiwan's indigenous cultures, revealing the complexities of cultural diversity and cultural transmission.

Research has found that the pattern of weaving also follows a formulaic pattern rule, especially in the NO.22 Klesan (men's and women's shawls), NO.61 Musha (men's and women's shawls), and NO.70 Truku group (men's and women's shawls) and other textiles. The structure is mainly derived from diamond patterns, and the weaving process is through modular substitution, constructing different shapes. The textiles show a composite stripe composition method with a horizontal separation structure, which is common in the sub-group textiles. The horizontal repetitive structure is secondary. In the composite stripes, the repetitiveness of the unit pattern does indeed exist in the arrangement rules, and the arrangement methods are divided into reflection and repetition. The combination of composite stripes can be divided into two colors, three colors, four colors, five colors, six colors, and seven colors, and the seven-color combination of composite stripes only appears in the Bunsui and the Beishi textiles. The use of color in the structure of composite stripes and the color configuration of the pattern boundary has no definite standard; most follow a dark and light color arrangement. Comparing the geographical distribution of sub-groups and the correspondence of textile patterns, some corresponding conjectures can be obtained from the interactive state produced by the migration process of some sub-groups. The unity or change rule of textile patterns in the traditional costumes of the Atayal tribe also highlights its inheritance and development issues. It is speculated that the reason may be to maintain the overall image of the sub-group costumes and to avoid similar patterns, so they must constantly seek changes, so the cultural meaning of the pattern can be regarded as the fusion of logical structure and ethnic emotions.

Contents

About the Authors

Po Hsun Wang is an associate professor in Design Science, the director of the Ph.D. program, and the vice dean at the Faculty of Innovation and Design, City University of Macau, China. He pursued his Ph.D. in Design Science from the National Yunlin University of Science and Technology, Taiwan, in 2015. He has published three monographs with Tsinghua University Press, China Business Press, and Economic Science Press and more than ten peer-reviewed articles. His research interests are visual communication design, design culture research, urban and rural planning and design, world cultural heritage, overall community construction, artistic and creative industries, visual communication design, city brand image, advertising, and marketing planning.

Jie Li is an assistant product design and development professor at the Faculty of Innovation and Design, City University of Macau, China. She gained her Doctor of Science degree from Aalto University, Finland, in 2022. She is responsible for a few research projects, such as Empathic Engineering, DigiTally, Tacit Knowledge Transformation, and User Ontology Model. She is also an invited reviewer of some essential design journals, such as the Journal of Mechanical Design, Journal of Engineering Design, Design Studies, and Design Science. Her research interests include human-centered design, user understanding, empathy, cultural influences in design, data-driven design, and user ontology models.

Chapter 1
Introduction

1.1 Traditional Clothing Culture of Taiwan's Minority Ethnic Groups

In the lives of minority ethnic groups in Taiwan, both spiritually and materially, they possess diverse characteristics, and the history of these ethnic groups has long been a focus of societal attention. In the past, Taiwan's minority ethnic groups were often referred to as indigenous people, called mountain tribes or Takasago tribes, and are currently collectively referred to as aborigines in Taiwan. In addition to the general term aborigines, there are also the Plains tribes, which are highly sinicized. Eskildsen mentioned that by 1869, the minority ethnic groups in southern Taiwan were gradually influenced by sinicization, especially in clothing and adornment, where Han Chinese styles emerged. The Plains tribes were named as such because they lived in the plains, adjacent to the Han Chinese, and over time, they integrated with the Han Chinese, leading to complete sinicization [1].

Furthermore, to distinguish the ethnic minorities living in the mountains, the ethnic minorities referred to in this study are not targeted at the Pepohoan. The culture and life of Taiwan's ethnic minorities have changed dramatically over time and have been influenced at various levels, differing significantly from their past appearances. Ethnic minorities originally possessed rich and diverse cultural content, whether in language systems, material culture, and social organization or even in the lifestyle of the ethnic group, all of which bear many characteristics of the Austronesian ethnic group to a considerable extent. These characteristics have led to many discussions about the movement of the Austronesian people. Taiwan's ethnic minorities have unique characteristics and are an important intangible cultural heritage. They have been the focus of attention for a long time, and from 1895 to 1945, a considerable amount of research has been accumulated and continues to grow.

The traditional costumes of Taiwan's ethnic minorities vary between groups, each with different cultural symbols and symbolic meanings. These differences are formed through dyeing and weaving patterns, which may have evolved through historical

P. H. Wang and J. Li, *Interpreting the Evolution of Atayal Tribal Textile Patterns*,
https://doi.org/10.1007/978-981-96-5451-2_1

changes or ethnic exchanges, permeating and merging, thus becoming a feature of ethnic distribution. Feroli mentioned that the Atayal tribe was once an enemy of the Amis and Bunun tribes. However, some Atayal villages still maintain trade relations with several Amis villages, and some Atayal people abandon their areas to live with the Bunun people [2]. From this, the cohabitation of different ethnic groups can easily lead to the exchange and integration of other cultures. Through the observation of ethnographic objects, material culture can also provide information to identify cultural contexts to a considerable extent. The early clothing materials of ethnic minorities were mainly derived from nature, whether fiber or color; some were obtained through exchange. The dyeing agents to produce ethnic minority clothing are all extracted from natural colors, and after the dyeing process, they are woven into cloth. In the process of dyeing and weaving, the combination of color and weaving methods expresses the unique imagery of each ethnic group. Due to the wide distribution of ethnic groups and the variety of woven patterns, Atayal woven products' pattern shapes and color combinations are also more diverse. Lisa Li mentioned that the clothing system of the Atayal tribe is a "square clothing system" (two narrow long strips of hemp cloth are folded and stitched into a collarless, sleeveless long dress or vest), and the composition of the square clothing is also a typical style among Taiwan's ethnic minorities [3].

The Atayal tribe is widely distributed and speculated to have developed earlier in Taiwan. The Atayal tribe is currently the second-largest ethnic group among Taiwan's ethnic minorities. From an anthropological perspective, the Atayal tribe is divided into two primary language sub-tribes. One is the Tayal sub-tribe (Tayal). The other is the Sediq sub-tribe (Sediq) [4, 5]. They are distributed throughout the north, central, and eastern regions of Taiwan, starting from Wulai Township in Taipei County (now renamed as Wulai District, New Taipei City), south to Xiulin Township, Zhuoxi Township, Wanrong Township in Hualien County, etc., with as many as 26 subgroups. The residential areas include Taipei County (now restructured as New Taipei City), Taoyuan County (now restructured as Taoyuan City), Hsinchu County, Miaoli County, Taichung County (now merged into Taichung City), Nantou County, Hualien County, Yilan County, etc., covering a wide area; the overlapping of sub-ethnic groups, coupled with the mobility of migration, marriage and other relationships, makes it challenging to clarify the characteristics or differences of clothing fabrics. Although the meaning of the textile patterns is unclear, it can be inferred that the visual symbols of clothing (weaving patterns, patterns) can serve as clues for ethnic identification.

The decorative patterns of Taiwan's minority ethnic clothing are woven, with both figurative and abstract patterns, and the constituent elements are primarily based on abstract geometric basic shapes, forming a diverse compound stripe. Abstract geometric forms are also common in other primitive arts, such as the objects of China's minority ethnic groups, where geometric pattern expressions can be seen. However, in the development of design art (craft art), the primary form of geometric patterns is an essential premise for learning form design, such as Constructivism born in Russia; the abstract geometric form of Bauhaus design education in Germany; Neoplasticism of the Dutch De Stijl movement, etc., all are based on geometric forms, constructing their visual design expression themes. Geometric forms, whether in ethnic decoration or modern design, are evident in various aspects of life objects, and the geometric patterns of minority ethnic-cultural symbols have formal aesthetic and particular symbolic implications and are the expression methods of early human decorative art.

The pattern structure of textiles is like a rational pattern composition. Textiles are woven by warp and weft intersecting at right angles. If the weft crosses the warp in a regular order of up and down, it is plain weave. The weft will form irregular patterns if it skips over two to three warps. Controlling the precise position of these patterns, creating a row and column, can weave a complete pattern. Patterns usually comprise two basic units arranged in a dot matrix calculated precisely by warp and weft, a basic binary system enriched by dyeing and other techniques to enhance its decorative effect. The pattern form of Atayal textiles shows that although the decorative patterns of different clothing forms have differences, these geometric patterns have a systematic phenomenon, and there may be a rigorous logical relationship between them. Given this, this study will interpret the pattern form and color expression on the clothing through the comparison and analysis of Atayal textiles and further understand the material culture of the Atayal.

It is necessary to explore the diverse cultural heritage and historical evolution of Taiwan's minority ethnic groups, with a particular focus on the Atayal tribe and their traditional textiles, to illustrate the significance of material culture in expressing ethnic identity.

1.2 Characteristics of Atayal Ethnic Group's Textile Clothing in Taiwan

The research results from 1895 to 1945, such as "Survey Report on Aboriginal Customs: Volume 1 Atayal," "Aboriginal Survey Report: Volume 7 Atayal Part 2, Volume 4 Seediq and Truku, Volume 5 Atayal Part 1″ [6, 7]. For instance, most minority ethnic groups have been found to engage in weaving activities, such as the Amis, Atayal, Bunun, Saisiyat, Shao, and Tao tribes, which are also a part of the daily work of minority women. Mackay mentioned in his research that minority women in Taiwan must learn household chores before marriage, including sewing and embroidering ceremonial dresses, using both manual and natural methods [8]. Japanese scholars have also pointed out in their research reports that the Atayal tribe is the most skilled at weaving, with textiles, bamboo, and rattan weaving as their main crafts. However, after 1945, weaving technology was gradually replaced by modern mechanical production, leading to the decline of traditional weaving among minority ethnic groups [9, 10]. Nowadays, in traditional tribes, weavers are in the minority. As time passes, the continuation of weaving skills also faces the dilemma of a decreasing number of weavers.

Previous research records show that each minority ethnic group has different dyeing and weaving expressions. The dyeing and weaving patterns of each tribe, through historical changes and interpretations of various life backgrounds, are also endowed with deeper symbolic meanings, such as expressing tribal styles, natural phenomena, historical stories, and even legends of racial origins. The textiles of the Atayal tribe are primarily based on diamond patterns and stripes, which are varied. For the Atayal tribe, the diamond pattern symbolizes the eyes of the ancestors and has a protective function. The pestle and mortar used for pounding or milling rice also often appear in the textile patterns, forming a characteristic feature. These patterns are grouped or repeated, showing the unique style of Atayal textiles. Redford and David Bright believe that any textile contains some specific primary weaving method, complex weaving patterns evolved from the constant repetition and combination of the minor symmetrical patterns, and using the "module combination" sequence can produce complex and varied structural forms [11, 12]. This research topic was developed based on this concept.

This study focuses on the content analysis of Atayal textiles, and the investigation of textile patterns and colors can construct a more concrete image of intangible cultural heritage. It can also serve as a necessary resource for developing the cultural industry. On the other hand, the current society's attention and assistance to Taiwan's minority ethnic groups allow the cultural heritage of these groups to continue and develop, such as crafts, performing arts, language, education, and other related cultural characteristics, which collectively become the driving force for sustainable cultural development, indirectly safeguarding the status of minority ethnic groups in modern society. Moreover, intangible cultural heritage is an indispensable element for industrial development. Therefore, this study has both academic and practical value. The objectives of the study are as follows:

1) The structure, form, and color relationships and characteristics of Atayal weaving are systematically analyzed, and patterns' similarities and differences are compared through subgroup distribution. Meanwhile, by researching the characteristics of the textile patterns, their composition rules are deciphered, and the module system of Atayal textile patterns is explored.
2) Understanding the characteristics of Atayal tribal textile patterns and colors can reveal the logical relationships inherent in pattern formation, providing the possibility to develop basic applications in visual design. By expanding the creative form through weaving techniques, different design paths can be opened, continuing the connection between Atayal tribal textiles and modern life.
3) Based on the anthropological surveys conducted by scholars in the Taiwan region from 1895 to 1945, such as the viewpoints discussed in extended linguistic anthropology on the origins of legends and tribal migrations, the characteristics and relationships between Atayal tribal textiles can be compiled, interpreting the fluid relationship between Atayal tribal textile patterns and sub-tribal groups, which can serve as corroborative evidence for understanding tribal movements.

4) For creators engaged in design professions or fine arts and crafts, this study can be used to depict the detailed structure of textile patterns and master the design expression of the Atayal tribal style.

1.3 Distribution of the Pan-Atayal Group and Textile Collections

Museums or private individuals mainly collected early Atayal tribal textiles, and it is now difficult to see traditional Atayal costumes in the tribes. Currently, the institutions in Taiwan that collect traditional costumes and artifacts of ethnic minorities, such as the Taiwan Museum[1] and the Anthropology Museum of the Department of Anthropology at Taiwan University[2] (from now on referred to as the Taiwan University Anthropology Museum), are relatively wealthy. The textiles in the collections are mainly based on cultural relics collected from ethnic minority tribes by anthropologists, such as Mori Ushinosuke[3] and Inō Kanori,[4] between 1895 and 1945 [13, 14].

1 The Taiwan Museum was established in 1908.

2 The Department of Anthropology at Taiwan University's Anthropology Museum (formerly the Specimen Room of the Folklore and Anthropology Lecture at Taipei University) was established in 1928.

3 Mori Chūnosuke (1877–1926) came to Taiwan in 1896 to conduct a "barbarian land" survey and followed anthropologist Torii Ryūzō into the mountains for investigation multiple times. In 1908, when the Taiwan Museum was established, Mori oversaw the "barbarian tribe" section of the exhibition and provided the cultural relics he collected for the museum's collection. The initial collection of "barbarian items" in the Taiwan Museum was almost entirely established by Mori alone. Due to Mori's pioneering achievements in investigating Taiwan's barbarian lands, he was often called the "Taiwan Barbarian Expert." Quoted from Huang Xuanwei: "Footprints of Anthropologists: A Centennial Special Exhibition of Taiwan Anthropology," published by the Museum of the Institute of Ethnology, Academia Sinica, Taiwan, 2011, page 34.

4 Inō Kanori (1867–1925), from 1895 to 1906, intermittently conducted field investigations and research on the social culture of Taiwan's minority ethnic groups. Although his original purpose was only for related policy planning, he paid considerable attention to specific material cultural relics in his fieldwork, such as house structures, room distribution, utensils, clothing, and ceremonial tools. He collected as many specimens as possible to corroborate abstract cultural meanings. When he left Taiwan in 1906, he also took them back to Japan. After Inō passed away due to illness in September 1925, his family agreed to donate this batch of specimens to the Folklore and Anthropology Lecture of the Department of Literature and Politics at Taiwan University,

Later, other cultural relics were collected from tribes during field investigations by Taiwanese anthropologists or purchased from private collectors. The history of these cultural relics is roughly around 1895–1945. Due to their age, the Taiwan Museum and the Taiwan University Anthropology Museum collections have historical significance and are of greater importance for material culture research. Therefore, the collections of these museums are listed as necessary sources for sample collection.

The Taichung County Cultural Centre (now renamed the Taichung City Government Cultural Bureau Huludun Cultural Centre) has been preserving, researching, promoting, and publishing Atayal weaving skills since 1990. It is also one of the institutions that possess Atayal cultural relics. According to the description in "Taichung County Cultural Centre Commissioned Collection of Atayal Weaving Cultural Relics Research: Compilation of Basic Information Cards for Cultural Relics Registration," the textiles in the collection were made about 50 years ago [15]. The content of the cultural relics registration is recorded based on the oral accounts of the original owners, as well as the "Survey Report on Aboriginal Customs: Volume One Atayal" [16] and "The Atayal People of Klesan: Ethnological Field Investigation and Research Volume One, Volume Two" [17] these research works serve as the basis for verification. The Museum of Prehistoric Taiwan also has a collection of cultural relics from some of Taiwan's minority ethnic groups. Although the museum is primarily for archaeological research, it also collects, researches, and educates material culture related to minority ethnic groups. Since 1992, it has systematically collected everyday items from Taiwan's minority ethnic groups, including a wide range of traditional costumes. Therefore, this study also includes the above institutions as subjects of investigation.

To diversify the sources of research samples, I specifically went to the Tenri University-affiliated Tenri Reference Museum in Japan, the Shung Ye Museum of Formosan Aborigines, and the Showakan Museum of Ancient Art (a private collection unit) to research the textiles of the Atayal tribe. The Tenri Reference Museum, established in 1930, houses some traditional costumes of Taiwan's minority ethnic groups collected between approximately 1966 and 1979. As the collection period is earlier than that of the Taichung County Cultural Centre, these samples can also provide insight into the characteristics of Atayal costumes at the time. The Shung Ye Museum of Formosan Aborigines, established in 1994, is the first private museum in Taiwan with a theme of minority ethnic groups. It houses some costumes for minority ethnic groups and is also one of the units from which this study draws materials. The Showakan Museum of Ancient Art, located in Fenyuan Township, Changhua County, has a collection of minority ethnic group artifacts that have been carefully collected over three generations. The operators also often purchase Taiwanese artifacts from abroad, especially those that have returned from Japan between 1895 and 1945. According to the person in charge, the historical background of the minority

which became the basic research equipment of the classroom. Quoted from Hu Jiayu and Cui Yilan: "Research on Inō Collection of the Department of Anthropology at Taiwan University," published by Taiwan University Press, 1998, page 24.

ethnic group artifacts in the museum is roughly between 1895 and 1945. The artifacts collected privately make it possible to compare their differences from various perspectives.

Although the research samples provided by the above institutions do not allow us to know the detailed weaving or collection years, the Atayal textiles obtained from different channels can provide a broad understanding of the style and characteristics of traditional Atayal costumes. The Atayal tribe, previously referred to as the Tayal and the Seediq sub-tribe, is now divided into the Truku tribe and the Seediq tribe, recognized as the 12th and 14th minority ethnic groups in Taiwan. Museums in Taiwan have separated some of their collections into Atayal, Truku, and Seediq, but some collections still need further verification. Although these tribes were considered the same in the past, and the patterns and colors of their textiles differ, the cultural context of the tribes is very similar. If we classify them according to the current tribal divisions, we may overlook the connections between the artifacts. The collection of Atayal textiles also includes traditional costumes of the Truku and Seediq tribes. In the process of classifying research samples, we compare them with related literature, such as "Research on the Affiliation of Taiwan's Minority Ethnic Groups: Volume 1 Main Text, Volume 2 Data Volume"[18], "Investigation of the Takasago Tribe: Overview of Barbarian Societies"[19], etc., and find that continuing the Nine Tribes Theory[5] and using barbarian societies and subgroups as the basis for classification is more appropriate [20]. In this way, we can also comprehensively examine whether there is a relationship between the style of textiles and geographical distribution.

According to anthropological rules for classifying ethnic groups, they are usually named based on their word for "human," supplemented by investigations into language, physique, and customs. Walis Norgan and Yu Guanghong mentioned in their book "History of Taiwan's Minority Ethnic Groups: The Atayal" that from the perspective of traditional social culture, the Truku and Seediq, along with the Tseole and Saikoliek, are all inseparable parts of the Atayal culture [21]. Sun Dachuan believes that language changes and geographical separation have led to communication difficulties, causing divergence in the Truku tribe's recognition of the Atayal name [22]. Lin Xiuche points out that driven by ethnic self-awareness based on self-naming, ethnic migration, tribal origins, and language communication difficulties, the Seediq people in the east and west have naturally left the Atayal group [23]. Due to the geographical distribution of the Atayal subgroups, influenced by tribal migration

5

The classification of Taiwan's minority ethnic groups, from 1895 to 1945, had so-called seven-part, nine-part, and twelve-part methods. The seven-part method includes the Rukai and Puyuma tribes in the Paiwan tribe. The nine-part method separates the two tribes above from the Paiwan tribe. The twelve-part method is determined from a linguistic perspective. From 1895 to 1945, the Shao tribe was positioned in the Pepohoan. Quoted from Tong Chunfa: "Takasago Chunqiu: The Culture and Art of Taiwan's Minority Ethnic Groups," Taiwan Lihong Publishing House 1996 edition, page 45.

or neighboring tribes, different cultural appearances have emerged, such as the significant differences in clothing styles between the Beishi and Klesans. The possibility of cultural overlap caused by tribal movement may lead to a composite phenomenon after cultural flow. Therefore, this study still regards the Truku and Seediq as part of the broader Atayal group, considering material cultural expressions, which may have their context, or exploring the connections between Atayal subgroups from the ethnic styles presented in material culture. Liu Qiwei also believes that the origin of an ethnic group can sometimes be observed materially [24]. Through the traditional costumes of the Atayal subgroups, the relationship between textile patterns and subgroup movement can be inferred by comparing the paths of subgroup movement.

Regarding the discussion of the origin and spread of the Atayal, many scholars currently speculate from different perspectives, such as language, mythology, history, and geography [25–27]. Still, all divide the Atayal into three major groups: the Saikoliek, Tseole, and Seediq. Taiwan's minority ethnic groups originally had no concept of an entire ethnic group but were based on communities or community units. Ethnic group classification is usually needed by scholars or administrators, from settlements, villages, communities, and ethnic groups to the Pepohoan and Gaoshan, and the formation of Taiwan's minority ethnic consciousness can be said to be recent events. This also shows that the actual boundaries of ethnic groups are not fixed, and what is seen at different stages is essentially just a hypothetical stable state at a cross-section of time [28]. There are many ways to classify the Atayal subgroups, most of which inherit the classification basis of anthropologists. The most common is based on the framework of "Research on the System of Taiwan's Gaoshan Tribe" by Nenozō Utsurikawa [29]. Today's academic classification system, commonly used for the Atayal group, essentially adopts the classification method inherited by Liao Shouchen from Nenozō Utsurikawa's collection [30–32]. This classification considers factors such as language, ethnic group, and region. The hierarchical relationship between each subgroup can be understood through this classification system.

While collecting artifact specimens, some samples could not be obtained for reasons such as the museum's preservation of older or damaged textiles, so they are closed to researchers for reference. According to research needs and scope, they can only be accessed through catalogs or digital collection databases. This is also one of the limitations of research investigation. The principles of selecting research samples include the background information of the textiles (era, collection location, affiliated tribe, subordinate group), pattern features, types of clothing, etc., while excluding those with unclear background information, repetitive textile styles, and severe appearance damage, to take textiles with more significant differences as research samples. The number of research samples is not absolute; it is mainly based on the source of the samples. The background information allows the Atayal textiles to be assigned to their respective subgroups. Among the museum and private collections, the Atayal textiles can be classified into the Tayal Saikolok group's Hakul (Hakku), Makanji, Msbtunux, Klesan, Malepa, Taranan; Tseole group's Klesan, Plnawan, Bunsui, Beishi, Nanshi, Meicyubus. Seediq sub-tribes Tkdaya (Musha and Pulibau) and Truku (Toroko). Therefore, the textiles of these subgroups are defined as the scope of research, as shown in Figs. 1.1 [33] and 1.2 [34].

Fig. 1.1 Atayal subgroup classification system and geographical distribution (Redrawn from Liao Shouchen, 1987) (Description: The Atayal people primarily inhabit the northern regions of Taiwan's Central Mountain Range, making them one of Taiwan's most widely distributed indigenous groups. Figure 1.1 showcases the classification system and geographical distribution of Atayal subgroups, including the specific locations of the subgroups represented in the study samples.)

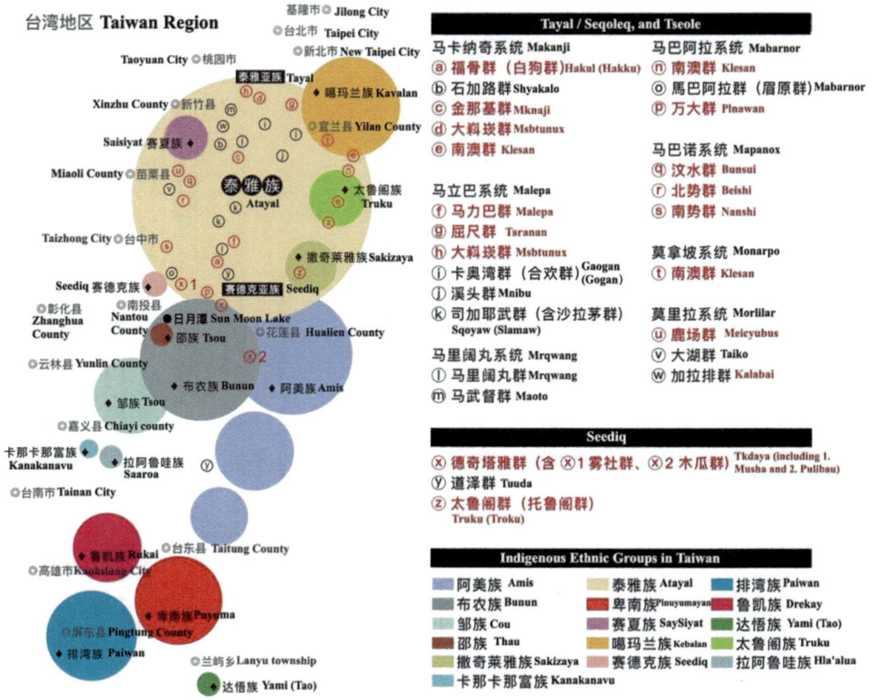

Fig. 1.2 Geographical distribution of Atayal subgroups (drawn by this study) (Description: Fig. 1.2 illustrates the geographical distribution of Atayal subgroups. The map provides a clear spatial representation, allowing viewers to understand the relative locations of each subset within the region.)

The data source of this study first uses the catalog published by the museum as a basis, including "*Taiwanese Minority Textiles: Weave Structure and Pattern Analysis*" [35], "*Museum Collections Atayal Weaving Artefacts Research: Compilation of Basic Information Cards for Artefact Registration*" [36], "*Clothing of Taiwanese Minorities: Collection of Tenri University Affiliated Tenri Reference Museum*" [37], "*Catalogue of Artifacts from the Shung Ye Museum of Formosan Aborigines*" [38], "*Research on the Yen Collection in the Department of Anthropology, National Taiwan University*" [39], "*Collection of the Museum of Prehistoric Culture in Taiwan: Catalogue of Minority Ethnic Textiles and Ornaments*" [40] etc. Suppose the museum has constructed a digital collection database. In that case, it can be used to further search for other samples available for research, such as the Museum of Anthropology at the National Taiwan University, Taiwan Museum, etc. The artifacts owned by the museum may not all be listed in the catalog, but the textiles presented in the catalog have their representative significance. By further inquiring about other textiles through the catalog content, the relationship between textiles can also be established. The number of samples held by the museum (sorted by research survey time) is explained as follows:

Taichung County Cultural Centre: The museum holds 108 Atayal textiles. Seventy-one pieces were purchased from Chan Hsiu-Mei, all belonging to the Huan Mountain tribe in He Ping Township, Taichung County. Judging from their appearance, material, and age, they are late-period textiles, including 34 pieces of cloth. Considering the diversity of the samples, these were excluded from the research samples. Another 37 pieces were purchased from Huang Shun-Yi and Huang Bai-Xing. These textiles are older, have a more comprehensive collection range, and have a more diverse type of textiles, so 17 pieces were selected as research samples.

Showakan Antique Art: A batch of Taiwanese minority ethnic textiles from around 1895 to 1945 was purchased from Japan, totaling 20 pieces. Among them, three pieces can be identified as Atayal textiles. Due to the unique pattern of the textiles, they were also included in the research samples.

Tenri University Affiliated Tenri Reference Museum: According to the museum catalog, there are 17 Atayal textiles. Excluding two pieces of beadwork clothing and bell clothing without patterns and deleting repeated styles of textiles, the total number of samples available for selection is 10.

Shung Ye Museum of Formosan Aborigines: The museum's collection primarily consists of utensils. According to the museum catalog, there are 19 Atayal textiles. Excluding four pieces of beadwork clothing, bell clothing, and women's long skirts without patterns, and deleting repeated styles of textiles and the number of samples not open for investigation, only one piece of fabric was selected as a research sample.

Museum of Anthropology, National Taiwan University: According to the specimen database, between 1928 and 1945, there are 69 Atayal textiles. Excluding 13 pieces of beadwork clothing and textiles without patterns, 13 pieces with unclear background information, and no collection location, there are 43 pieces left. After deleting repeated styles of textiles and the number of samples not open for investigation, the total number of samples available for selection is 19.

Taiwan Museum: According to the digital collection database, there are 321 Atayal textiles. Excluding 83 pieces of beadwork clothing, bell clothing, and textiles without patterns, 175 pieces with unclear background information and no collection location, and eight pieces that cannot be confirmed as Atayal textiles, are 55 pieces left. After deleting repeated styles of textiles and the number of samples not open for investigation, the total number of samples available for selection is 15. In addition, pieces like Atayal textiles were explicitly included, illustrating the difficulties encountered when investigating artifact specimens that cannot be categorized.

Taiwan Prehistoric Culture Museum: According to the museum's catalog, there are 22 pieces of Atayal tribal textiles. There are nine pieces, excluding the shell bead clothing and patternless textiles, leaving 13. After removing duplicate styles of textiles, the total number of samples available for selection is 8.

Based on the established research motivation, purpose, and scope, this research framework is designed and developed, as shown in Fig. 1.3 outlines the core issues

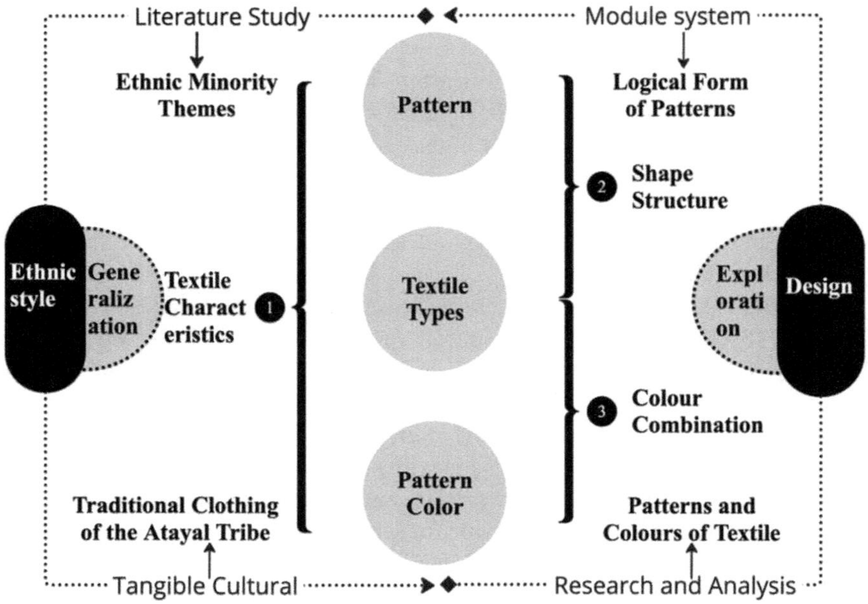

Fig. 1.3 Research Framework (Drawn by this study) (Description: Fig. 1.3 presents the research framework, focusing on the forms and colors of Atayal textile patterns. These textiles symbolize the Atayal people's cultural identity and reflect the distinctive characteristics of their traditional craftsmanship)

this study aims to explore. Ethnic objects result from production and manufacturing; the production method is the most basic cultural phenomenon. Under certain conditions, the decorative form on the object's surface is artistic. The expression of traditional costumes of Taiwan's minority ethnic groups implies decorative thinking in the development process of primitive art, and it contains the technical and logical relationship of craft production and pattern design. For the Atayal tribe, the shape and color of the textile patterns symbolize the cultural characteristics of the tribe. Whether it is the structural form or the color combination, to some extent, there is a design context to follow, which is also the unique clothing style of the ethnic group. This study aims to clarify the structural form and color combination of the patterns, and another focus is to explore the design thinking of the woven patterns, summarize the various styles of textile patterns, and even construct the possibility of its module system through research and analysis.

References

1. Robert Eskildsen, "*Foreign Adventurers and the Aborigines of Southern Taiwan", 1867–1874* (Taipei: Institute of Taiwan History, Academia Sinica of Taiwan, 2005), P.157.

2. Fei Luo Li: "*Exploration of the Culture and Language Classification of Taiwan's Aboriginal Tribes*," Institute of Ethnology, Academia Sinica, Taiwan, 1994 edition, page 435.
3. Lisa Li: "*Taiwanese Minority Ethnic Clothing Culture: Tradition, Meaning, Illustration*," Nantian Bookstore Co., Ltd., Taiwan, 1998 edition, page 81.
4. Folklore and Anthropology Research Department, Taipei University: "*Research on the System of Taiwanese Minority Ethnic Groups: Volume One*," translated by Yang Nanjun, Taiwanese Minority Ethnic Committee, Nantian Bookstore Co., Ltd., Taiwan, 2011 edition.
5. Li Rengui: "*History of Taiwanese Minority Ethnic Groups: Language Volume*," Taiwan Literature Museum, 1999 edition.
6. Taiwan Old Custom Survey Association: "*Aboriginal Custom Survey Report: Volume One, Atayal Tribe*," translated by Taiwan Central Research Institute of Ethnology, Taiwan Central Research Institute of Ethnology, 1996 edition.
7. Taiwan Old Custom Survey Association: "*Aboriginal Survey Report: Volume Seven, Atayal Tribe Part Two, Volume Four, Seediq Tribe and Truku Tribe, Volume Five, Atayal Tribe Part One*," translated by Taiwan Central Research Institute of Ethnology, Taiwan Central Research Institute of Ethnology, 2010, 2011, 2012 editions.
8. George Leslie Mackay: "*From Far Formosa: The island, its people and missions*," (Taipei: Ch'eng wen publishing company, 1972), P.299.
9. Mori Ushinosuke: "*Taiwan Aboriginal Record: Volume One*," Nantian Bookstore Co., Ltd., Taiwan, 1917 edition.
10. Sato Bunichi: "*Research on the Primitive Art of Taiwanese Indigenous Peoples*," Nantian Bookstore Co., Ltd., Taiwan, 1988 edition.
11. Redford: "*Wanwu: Modularisation and Mass Production in Chinese Art*," translated by Zhang Zong, Sanlian Bookstore, 2006 edition, pages 4–11.
12. David Brett: "*New Thinking in Decoration: Pleasure and Ideology in Visual Arts*," translated by Zhang Hui, Tian Lijuan, Wang Chunchen, Jiangsu Fine Arts Publishing House, 2006 edition, pages 284–285.
13. Huang Xuanwei: "*Footprints of Anthropologists: A Centennial Special Exhibition of Taiwan Anthropology*," published by the Museum of the Institute of Ethnology, Academia Sinica, Taiwan, 2011.
14. Hu Jiayu and Cui Yilan: "*Research on Inō Collection of the Department of Anthropology at Taiwan University*," Taiwan University Press, 1998.
15. Taichung County Cultural Centre: "*Taichung County Cultural Centre Commissioned Museum Collection of Atayal Weaving Cultural Relics Research: Compilation of Basic Data Cards for Cultural Relics Registration*," published by Taichung County Cultural Centre, Taiwan, 2001.
16. Taiwan Old Custom Survey Association: "*Report on the Survey of Aboriginal Customs: Volume One, Atayal*," translated by the Institute of Ethnology, Academia Sinica, Taiwan, published by the Institute of Ethnology, Academia Sinica, Taiwan, 1996.
17. Li Yiyuan, Shi Lei, Ruan Changrui, Yang Fufa: "*The Atayal People of Klesan: Ethnographic Field Investigation and Research Volume One and Two*," published by the Institute of Ethnology, Academia Sinica, Taiwan, 1963 and 1964.
18. Taipei University Folklore and Anthropology Research Room: "*Research on the System Affiliation of Taiwan's Minority Ethnic Groups: Volume One, Main Text, Volume Two, Data*," translated by Yang Nanjun, published by the Taiwan Minority Ethnic Committee and Nantian Bookstore Co., Ltd., Taiwan, 2011 and 2012.
19. Taiwan Police Bureau Savage Course: "*Takasago Tribe Investigation: Savage Society Overview*," translated by the Institute of Ethnology, Academia Sinica, Taiwan, 2011 edition of the Institute of Ethnology, Academia Sinica, Taiwan.
20. Tong Chunfa: "*Takasago Chunqiu: The Culture and Art of Taiwan's Minority Ethnic Groups*," Taiwan Lihong Publishing House 1996 edition, page 45.
21. Walis Nokan, Yu Guanghong: "*History of Taiwan's Minority Ethnic Groups: Atayal History*," Taiwan Literature Museum 2002 edition, page 6.
22. Sun Daquan: "*Inventory of Minority Ethnic Culture and Oral History in Truku Park*," commissioned research report by Truku Park Management Office, Taiwan, Taiwan Minority Ethnic Culture Development Association 2006 edition, page 22.

23. Lin Xiuche: "*Seediq Tribe Renaming,*" commissioned research by Taiwan Minority Ethnic Committee, Taiwan Law and Politics University Minority Ethnic Research Center 2007 edition, page 80.
24. Liu Qiwei: "*Primitive Art Exploration,*" Taiwan Qiyuan Bookstore Co., Ltd. 1993 edition, page 193.
25. Li Rengui: "*Taiwan Austronesian Ethnic Groups and Migration (Revised New Edition),*" Taiwan Vanguard Publishing House 2011 edition.
26. Pu Zhongcheng: "*The Forgotten Sanctuary: Tracing Minority Ethnic Mythology, History and Literature,*" Taiwan Wunan Book Publishing Co., Ltd. 2007 edition.
27. Taipei University Folklore and Anthropology Research Room: "*Research on the System of Taiwan's Minority Ethnic Groups: Volume One Main Text,*" translated by Yang Nanjun, Taiwan Minority Ethnic Committee, Taiwan Nantian Bookstore Co., Ltd. 2011 edition.
28. Liu Yichang: "*History of Taiwan's Minority Ethnic Groups: Prehistoric Volume,*" Taiwan Literature Museum 2002 edition, page 76.
29. Taipei University Folklore and Ethnology Research Room: "*Research on the System ofTaiwan's Minority Ethnic Groups: Volume One Text, Volume Two Data,*" translated by Yang Nanjun, Taiwan Minority Ethnic Groups Committee, Taiwan Nantian Bookstore Co., Ltd., 2011, 2012 edition.
30. Ma Tengyue: "*Atayal Tribe Facial Tattoo Atlas,*" Taiwan Photography World Magazine, 1998 edition.
31. Wang Meixia: "*Atayal Tribe,*" Taiwan Sanmin Book Co., Ltd., 2006 edition.
32. Kang Peide: "*The Beauty and Sorrow of the Atayal Tribe's msbtunux: The Family History of the KButa Clan,*" Taiwan Minority Ethnic Groups Committee, 2009 edition.
33. Liao Shouchen: "*The Culture of the Atayal Tribe: Tribal Migration and Expansion,*" Taiwan World News College, 1987, pages 8–9.
34. Ma Tengyue: "*Atayal Tribe Facial Tattoo Atlas,*" Taiwan Photography World Magazine, 1998 edition, page 33.
35. Cai Yushan: "*Taiwan's Minority Ethnic Groups' Textiles: Analysis of Weaving Structure and Patterns,*" Taiwan Taichung County Cultural Centre, 2006 edition.
36. Jiang Shaoying: "*Research on Atayal Tribe's Woven Cultural Relics in the Collection: Compilation of Basic Data Cards for Cultural Relics Registration,*" Taiwan Taichung County Cultural Centre, Taiwan Taipei University of the Arts Traditional Art Research Institute, 2001 edition.
37. Tenri University: "*Costumes of Taiwan's Minority Ethnic Groups: Collection of Tenri University Affiliated Tenri Reference Museum,*" Tenri University, Japan, 1993 edition.
38. Xu Yahui: "*Catalogue of Cultural Relics in the Shunyi Taiwan Minority Ethnic Groups Museum,*" Shunyi Taiwan Minority Ethnic Groups Museum, 1999 edition.
39. Hu Jiayu, Cui Yilan: "*Research on the Ino Collection of the Department of Anthropology, National Taiwan University,*" National Taiwan University Press, 1998 edition.
40. Fang Junwei: "*Taiwan Prehistoric Culture Museum Collection: Catalogue of Minority Ethnic Textiles and Ornaments,*" Taiwan Prehistoric Culture Museum, 2002 edition.

Chapter 2
Taiwan's Minority Ethnic Groups and the Pan-Atayal Tribe

2.1 Minority Ethnic Groups and Cultural Connotations

2.1.1 Evolution of Minority Ethnic Group Classification

According to the literature records from 1895 to 1945, Taiwan's minority ethnic groups settled here much earlier than the Han people. Their living range and ethnic group definition were constructed in the research surveys of anthropologists, such as the seven tribes advocated by Inō Kanori and Kurino Hiroshi in 1898, the nine tribes advocated by Torii Ryūzō in 1901, the nine tribes classified in the Taiwan Aborigines Report in 1911; the six tribes advocated by Mori Ushinosuke from 1912 to 1917; the nine tribes advocated by Nenozō Utsurikawa in 1935; the eight tribes advocated by Kano Tadao in 1941 [1, 2]. The nine tribes were advocated by the Taiwan University Ethnology Research Room in 1948; the nine tribes were announced by the Taiwan Ministry of the Interior in 1954; and the nine tribes were announced by the Taiwan Minority Ethnic Committee in 1988. The previously recognized nine tribes included the Amis, Atayal, Paiwan, Bunun, Puyuma, Rukai, Tsou, Saisiyat, and Tao (Tau). Subsequently, the Thao was announced as the 10th tribe of Taiwan's minority ethnic groups in 2001; the Kavalan was officially recognized as the 11th tribe in 2002; the Truku was announced as the 12th tribe in 2004; the Sakizaya was announced as the 13th tribe in 2007; the Sediq was announced as the 14th tribe in 2008; the Hla'alua and Kanakanavu were successively announced as the 15th and 16th tribes from 2010 to 2011. Up to now, there are 16 recognized minority ethnic groups in Taiwan, which does not include the Pepohoan.

Based on the accounts in the literature mentioned above, some scholars argue that Taiwan's ethnic minorities should no longer be subdivided into Gaoshan and Pepohoans but should consider the differentiation of ethnic minorities, abandoning the standards of Sinicization and residential division. Instead, they should reclassify sub-ethnic groups more comprehensively and academically oriented from aspects such as bloodline, language, migration, and cultural typology [3]. In surveying Taiwan's

P. H. Wang and J. Li, *Interpreting the Evolution of Atayal Tribal Textile Patterns*,
https://doi.org/10.1007/978-981-96-5451-2_2

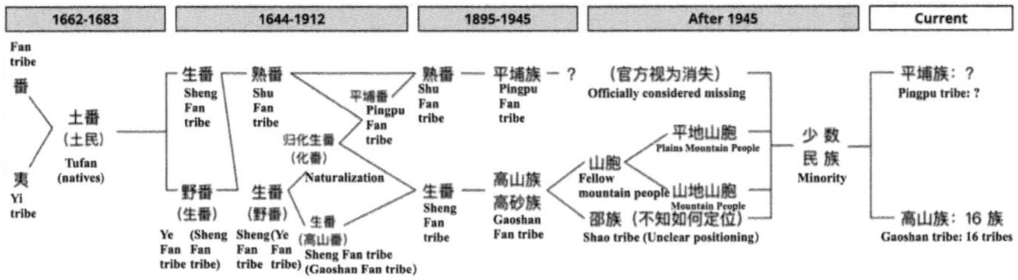

Fig. 2.1 Evolution of the classification of Taiwan's ethnic minorities (Redrawn from Xiao Qiongrui, 1999)

ethnic minority textiles, it is necessary first to review the evolution of these ethnic groups' names in a general way, as shown in Fig. 2.1. This will help clarify the relationships among the various ethnic groups of the minorities and further understand the differences in their cultural identities. However, the significance of cultural relics lies in the uniqueness presented by the culture behind the ethnic groups, which is enough to illustrate the importance of Taiwan's ethnic minorities that cannot be ignored, and it is worth further in-depth exploration in the future.

2.1.2 Characteristics of Ethnic Minority Art

Ethnic minority art is often referred to as primitive art or folk crafts. Morphy pointed out that ethnic minority art was discovered during the colonization process in Australia, and this "ethnic minority art" has become a vocabulary under the European colonial system [4]. Wang Songshan mentioned in his book "Contemporary Taiwanese Ethnic Minority Art" that because most people regard primitive art as simple, backward, and a more primitive form of artistic expression, anthropologists, in addition to emphasizing the descriptive nature of the term "primitive," also refer to non-Western cultures, especially tribal societies' artistic expressions, as ethnic art. The term "ethnic" is more important because art is seen as an essential content of the social and cultural category [5]. Forty-seven artistic expressions of Taiwan's ethnic minorities are not primitive art forms.

To some extent, they involve complex craft techniques, such as the wood carving crafts of the Paiwan and Rukai tribes, the weaving crafts of the Atayal tribe, etc., all of which combine craft techniques and aesthetic form. Jiang Xun discussed that the original meaning of "art" refers to craftsmanship and skills, and it initially occurred inseparably from manual work. In its early meaning, art is humans' understanding and utilization of specific material characteristics. Although it contains fundamental concepts, cognition, and thinking components, this understanding and utilization of materials is most closely related to human hands in their visible form. Therefore, this kind of art generally refers to the skill of the hand in controlling (making) a material [6]. Xu Gongming pointed out that in the language and concepts of Taiwan's ethnic minorities, there is no concept of craft integrated under the influence of modern Western aesthetics, nor is there an equivalent vocabulary for it, nor the word "artist" and there is no concept to distinguish between craft and art. Only words with similar meanings exist, such as vents in the Paiwan language or tiled in the Amis language, both of which mean patterns, designs, characters, or symbols [7].

The traditional arts of Taiwan's ethnic minorities are manifested in various aspects of life, including ceremonial rituals, craft production, pattern design, color preferences, clothing, architecture, housing, and musical instruments, all of which have different artistic achievements. Through the publication of works related to the study of ethnic minorities, such as "Taiwan's Ethnic Minorities Customs" [8], "Taiwan Culture Series: Taiwan's Ethnic Minorities" [9], "Taiwan's Ethnic Minorities Art Field Notes" [10], "Taiwan's Ethnic Minorities" [11], etc., by comparing the content of each document, one can understand the cultural characteristics of Taiwan's ethnic minorities. The discussion on the artistic characteristics of ethnic minorities is as follows:

1. Ceremonial rituals: Rituals are essential cultural activities of ethnic minorities and a reflection of life, symbolizing the meaning of generational inheritance. Regardless of birth, naming, adulthood marriage, and life rituals throughout one's life, all have special celebration ceremonies and inheritance significance. Because of farming, there are rituals for clearing, sowing, weeding, and harvesting; because of hunting, from going up the mountain to triumphant return, the tribe holds celebrations for the warriors; because of the remembrance of ancestral spirits, each tribe has unique ceremonies; because of food, clothing, housing, and transportation, the completion of weaving, food, building, and transportation tools, each step of the ritual is key. Ethnic minorities all have different and diverse rituals, such as the Amis' Harvest Festival, the Atayal's Harvest Festival and Ancestor Worship, the Paiwan's Five-Year Festival, the Bunun's Ear Shooting Festival, the Puyuma's Monkey Festival, and Big Hunt Festival; the Rukai's Millet Harvest Festival; the Tsou's War Festival; the Saisiyat's Pas-Tai Festival; the Yami's Flying Fish Festival; the Thao's Eel Worship and New Year Festival; the Kavalan's New Year Ancestor Worship; the Truku's Ancestor Worship, etc. These ceremonies represent respect for traditional culture and are a testament to the continuation of life through generations.
2. Craft production: In many ethnic cultures, creative artists and exceptional craft technicians (like those with extraordinary abilities) are often seen as having natural skills, or their artistic skills come from the bestowal of supernatural powers; even in some cultures, there are always deities in charge of similar creative abilities [12]. From the craft system of ethnic minorities, we can perceive the expression of primitive art, such as clothing, utensils, houses, etc., which have become representative craft symbols of various ethnic groups. Crafts such as bamboo and wooden spoons, textiles, pottery, woven goods, carvings, etc., also represent the production of ethnic minorities in response to the needs of life, environment, and customs. The Amis, Paiwan, Bunun, Puyuma, Rukai, Saisiyat, and Yami tribes' carvings, pottery, woven goods, and textiles; the Atayal

tribe's weaving; the Paiwan Rukai tribes' glass beads; the Paiwan, Bunun, Rukai, Tsou tribes' leather making; the Yami tribe's shipbuilding, blacksmithing, etc., each ethnic group has its specialized craft field, using different materials for the crafting process, also creating different ethnic styles under different cultural backgrounds.

3. Pattern formation: The creation and preservation of visual materials is an integral part of the overall record of human civilization. It is the starting point of civilization and an endless cycle of symbolic systems. Human visual records are the origin of symbols, and visual symbols are the starting point and sum of materialized civilization [13]. Through symbolic activities, humans create the sense and order of their world. Humans use words, objective objects, images, postures, etc., and all symbol skills to perform judgment and interpretation activities. They use concrete, specific, and secular things to explain universal, infinite, and sacred things [14]. The pattern formation of ethnic minorities has different styles, such as the plant patterns of the Amis tribe, the diamond patterns, the geometric patterns of the Atayal, Puyuma, Saisiyat, Thao, and Truku tribes, the sun patterns, and hundred-step snake patterns. Pottery patterns, human head patterns, and human figure patterns of the Paiwan Rukai tribes; the hundred-step snake diamond patterns of the Bunun tribe; the human dance patterns of the Puyuma tribe; the triangle square patterns of the Tsou tribe; the wave patterns, sun patterns, spiral patterns, human figure patterns of the Yami tribe, etc. Chen Qilu's investigation of the Paiwan group revealed that the headman class's house beams, wall panels, and wooden utensils are all carved with patterns. Human figures, heads, snakes, and deer are the m. The use of other animal patterns may be more recent. Geometric patterns are primarily decorative and may have evolved from human head and snake patterns [15]. With the snake as a totem,[1] each clan of the Saisiyat tribe has its totem mark, generally related to natural phenomena, animals, and plants [16]. Chen Yulan also mentioned that the most common pattern is that the "spirit snake" is the totem of the Saisiyat, Jie, and Xia clans [17]. Hong Yingsheng believes that the totem of the Rukai tribe's hundred-step snake, the more snakeheads, may indicate more substantial witchcraft power [18]. Although the sources of pattern themes between ethnic groups are similar, they still have different meanings and symbols worth exploring.
4. Color Preferences: Among ethnic minorities, preferences in color usage can be discerned from the woven or embroidered patterns of their clothing. For instance,

[1] Totems are a visual symbol in primitive societies, widely distributed in North America, the Pacific Islands, Australia, Africa, etc. The word "totem" originates from the Ojibway tribe language of North American Indians, originally meaning a kind of supernatural protector. The so-called totem is the awe of the natural world phenomena by people in primitive times due to their lack of knowledge. It is not an organized religion, but the awe and superstition of primitive humans towards the natural world; they regard a certain natural thing as having a special close relationship with their clan, use it as a totem and banner, as a common symbol of a social group, regard it as sacred and inviolable, and worship it. Generally, animals or plants are used as totems, and sometimes natural things and phenomena are used as totems, such as mountains, rivers, sun, moon, stars, wind, rain, thunder, lightning, etc. Quoted from Darcy Ula Bay · Bima: "Taiwan's Indigenous Culture Series: Taiwan's Minority Bunun Tribe," published by Taiwan Indigenous Publishing House, 2003 edition, page 81.

the Amis tribe predominantly uses black and dark blue; the Atayal and Truku tribes use black, red, blue, and white; the Paiwan, Puyuma, Rukai, Tsou, Saisiyat, Yami, and Thao tribes use black, red, and white; the Bunun tribe uses dark blue, etc. Ethnic minorities have also been influenced by Sinicisation and the proximity of their residences to each other, leading to mutual influences in clothing and color, such as the Amis tribe in the northern and central regions of Taiwan mainly wearing a type of skirt like a sword belt. In contrast, they wear a single skirt with embroidered patterns in the south. Women's clothing is mainly in red and black. The Bunun tribe has been using a lot of golden yellow to highlight their clothing in recent years. The Thao tribe's clothing is like that of the Tsou tribe. Thus, the color representation of Taiwan's ethnic minority clothing has changed.

The patterns, designs, or symbols that symbolize ethnic minority cultures can be seen as a unique ethnic style. In recent years, indigenous craft movements have been resurgent due to the boost from the cultural and creative industries. Many objects carry the visual culture of ethnic minorities through their manufacture, integrating different forms, colors, and materials, such as traditional clothing, daily utensils, and architectural facilities, becoming the commonly recognized image of ethnic minorities. Because of this, designs that imitate ethnic styles or other patterns in terms of form can achieve a certain degree of similarity. However, the relationship between symbols and objects and the hidden meanings of ethnic characteristics often lead to diametrically opposite phenomena far from the spirit of ethnic culture. Therefore, when using the unique visual image of ethnic minorities for translation design, more attention should be paid to connecting and expressing deeper meanings, not just borrowing symbolic forms.

2.1.3 Ethnic Minority Art Themes

Through the Taiwan Master's Thesis Knowledge Value System[2] and the Government Research Information System[3] websites, by entering keywords such as "aborigines" (ethnic minorities), "Atayal," "textiles," "traditional clothing," "patterns," "colors," etc., related research topics on Taiwan's ethnic minority art are searched and detailed as follows:

Ethnic studies and minority ethnic issues have always been the focus areas of Taiwan's humanities and social sciences. Most dissertations about Taiwan's minority ethnic art are concentrated in humanities and arts, with research topics including weaving crafts, traditional costumes, totem decorations, mythological stories, educational curriculums, applied design, etc. Research institutions include the Department of Fine Arts at Taiwan University of Arts, Institute of Plastic Arts; Institute of Ethnic

[2] Taiwan Master's Thesis Knowledge Value System, https://reurl.cc/aG6Wo4, 23rd October 2022.

[3] Taiwan Regional Government Research Information System, http://www.grb.gov.tw, 23rd October 2022.

Relations and Culture, Institute of Children's Literature at Donghua University; Institute of Museum Studies and Antiquities Conservation at Tainan University of the Arts; Institute of Design at Yunlin University of Science and Technology; Department of Commercial Design at Taichung Institute of Technology (now renamed Taichung University of Science and Technology); Institute of Tropical Agriculture and International Cooperation at Pingtung University of Science and Technology; Institute of Anthropology at Tsinghua University; Institute of Art at Cheng Kung University; Institute of Industrial Education at National Taiwan Normal University; Institute of Leisure Health Management at Tajen University of Science and Technology; Institute of Multimedia and Computer Entertainment Science at Southern Taiwan University of Science and Technology; Institute of Art at Fo Guang College of Humanities and Social Sciences (now renamed Fo Guang University); Department of Chinese Literature at Soochow University; Department of Textile and Clothing at Fu Jen Catholic University; Institute of Applied Design at Shu-Te University. Among them, it is worth noting that researchers from institutes such as the Institute of Ethnic Relations and Culture, Institute of Children's Literature at Donghua University, Institute of Museum Studies and Antiquities Conservation at Tainan University of the Arts, Institute of Tropical Agriculture and International Cooperation at Pingtung University of Science and Technology; Institute of Anthropology at Tsinghua University; Institute of Industrial Education at National Taiwan Normal University; Institute of Leisure Health Management at Tajen University of Science and Technology; Department of Chinese Literature at Soochow University, etc., are trying to discuss the artistic characteristics of minority ethnic groups from other academic fields. The topics related to this study, such as "Research on the Deconstruction and Reconstruction of Taiwan's Minority Ethnic Textiles" [19], "Research on the Diamond Pattern of the Atayal Tribe" [20], "Research on the Use of Symmetry in Minority Ethnic Costumes" [21], "Research on the Art of Atayal Weaving" [22], "Research on the Changes in Traditional Crafts of the Atayal Beishi" [23]. These studies mainly explore the weaving techniques of the Atayal tribe. Still, because most of the research samples use museum-published collection catalogs or modern Atayal tribe textiles collected during field investigations as samples, they have lost the authenticity of the original specimens based on the analysis of cultural relics. Even the collected cultural relics do not belong to the Atayal tribe in terms of their pattern characteristics, leading to deviations in research results. Current research on the art of Taiwan's minority ethnic groups, although differing in content and emphasis, lacks a further discussion on the patterns and colors of textiles, and the differences between patterns, colors, and ethnic groups have not yet been systematically summarized and analyzed. Therefore, this study attempts to discuss the system of ethnic patterns from the relationship between the pattern shaping and color expression of textiles, hoping that this study can make up for the omissions in the research on the artistic characteristics of Taiwan's minority ethnic groups.

The research methods mentioned above include field investigations (interviews, participatory observations), literature analysis, image analysis, comparative analysis, historical research, sample collection, focus groups, morphological analysis, stylistic history, questionnaire statistics, etc. The essential purpose of these research methods is to provide a standard model for inspection or analysis, and following the method procedures can understand the related questions in the study. For example, field investigation interviews (interviews, in-depth interviews), participatory observations are commonly used in interpreting contemporary cultural phenomena or life situations, such as "The Study of the Inheritance Education of Atayal Weaving Craft: Taking Fuxing Township, Taoyuan County as an Example"[24], "I Weave, Therefore I Am: The Cultural Production of Female Craft Weavers of the Truku Tribe in Xiulin Township, Hualien County" [25], "The Study of Deconstruction and Reconstruction of Taiwan's Minority Ethnic Textiles" [26], "The Study of Paiwan Dress Culture: Taking the Lavavul Group of Sandimen Tribe as an Example" [27], "The Study of Atayal Diamond Pattern" [28], "The Application of Taiwan's Minority Ethnic Traditional Totem Art to Wood Art Creative Design: Taking Taitung Beinan Tribe Totem as an Example" [29], "The Commercialization of Sandimen Paiwan Clothing and Its Meaning Change" [30], "The Study of Taiwan's Minority Ethnic Tattoo Myths and Legends" [31], "The Change and Formation of the Wedding Dress of the Rukai Tribe in Wutai: A Case Study of the Ba Family Wedding" [32], "New Styles of Paiwan Women's Clothing: An Exploration of Its Form and Content" [33], "A Study on the Origin of Shao Ethnic Group's Clothing" [34], "A Study on the Relationship between Bunun Ethnic Group's Textile Clothing and Identity: A Case Study of the Kadu Tribe in Zhuo Community" [35], "A Study on the Changes in Traditional Crafts of the Atayal Tribe in the North" [36], "A Study on the Weaving of the Saisiyat Tribe, a Minority Ethnic Group in Taiwan" [37], As the research content involves the history, culture, and life of the ethnic groups, the investigation and analysis of literature is also one of the methods often used in these research topics.

The method of image analysis mainly explores the research topics of material culture changes, such as "A Study on the Inheritance Education of Atayal Weaving Craft: A Case Study of Fuxing Township in Taoyuan County" [38], "A Study on the Clothing Culture of the Pepohoan: A Case Study of the Pepohoan's Clothing Artifacts in Taiwan Museum" [39], "A Study on the Use of Symmetry in Minority Ethnic Group's Clothing" [40], "A Study on the Myths and Legends of Tattooing in Taiwan's Minority Ethnic Groups" [41], "A Study on the Relationship between Bunun Ethnic Group's Textile Clothing and Identity: A Case Study of the Kadu Tribe in Zhuo Community" [42], In the study of "A Study on the Clothing Culture of the Pepohoan: A Case Study of the Pepohoan's Clothing Artifacts in Taiwan Museum" [43], it proposes a triangulation method based on text, image, and artifacts, which can verify the consistency of artifacts and literature analysis results. This method can provide more accurate background information on material culture, which is also the data comparison method expected to be adopted in this study to identify the sub-tribes of Atayal textiles.

Other research methods, such as comparative analysis, historical research, sample collection, focus groups, morphological analysis, stylistic history, questionnaire statistics, etc., are adopted to solve problems according to different research needs. Among them, comparative analysis, morphological analysis, and the research methods used in this study are similar, which can interpret the connotation of traditional materials, such as "A Study on the Deconstruction and Reconstruction of

Taiwan's Minority Ethnic Group's Textiles" [44], "A Study on the Diamond Patterns of the Atayal Tribe." [45], "Research on the Tattoo Myths and Legends of Taiwan's Minority Ethnic Groups" [46], "Study on the Changes and Formation of the Wedding Attire of the Rukai Tribe in Wutai: A Case Study of the Ba Family Wedding" [47], "Research on the Totem Stories of the Paiwan and Rukai Tribes" [48], but these research topics, when discussing material forms, only use text or images for analysis and description, or quote previous research results. The researchers should have invested in the research analysis process and needed detailed pattern descriptions. Although this process is time-consuming and complicated, it can grasp the profound cultural clues of the material. This is also a critical breakthrough in the methodological considerations of this study. As mentioned above, understanding research topics with the same attributes can help one understand the issues that other scholars focus on and the focus of these research topics from their research methods. By reflecting on this, future research analysis can appropriately select research methods for analysis and propose more specific research results.

By querying the Science and Technology Committee's research plan in the government research information system of the Taiwan region, research on related Taiwan minority ethnic art topics that have been approved and approved can be found. The research topics include pattern analysis, such as "Research on the Cultural Characteristics of Taiwan's Minority Ethnic Groups from the Colors and Patterns of Textiles (I, II)" (2011) (2009); "The Formation and Transformation of Rukai Patterns: A Case Study of the Qingye Tribe in Sandimen" (2009); "Cultural Communication Research of Minority Ethnic Handicrafts under the Context of Tourism" (2007); "Research on Tourists' Perception of the Authenticity of Minority Ethnic Tourism Cultural Products: A Case Study of the Tayal Tribe's Weaving Crafts in the Wulai Area" (2006). Visual images, such as "Research on the Visual Images of Taiwan's Minority Ethnic Groups: A Case Study of the Paiwan and Rukai Tribes' Myths, Legends and Ritual Ceremonies" (2011); "Research on Minority Ethnic Images: Photography, Memory and Local Perspectives" (2007); "Research and Application of Taiwan's Minority Ethnic Image Culture in Product Design" (2004). Digital collections, such as "Digital Collection and Learning's Overseas Promotion and International Cooperation Plan (Sub-Plan Three): Taiwan's Lost Overseas Museum Treasures Digitalisation Plan; Taiwan Ethnology Collection Cross-National Research and Exchange Plan; Royal Ontario Museum, Canada 'Taiwan Collection'" (2011); "Taiwan Museum's Minority Ethnic Cultural Relics Digitalisation Plan" (2007); "Taiwan's Minority Ethnic Textile Clothing Culture Digital Collection Creative Value-Added Plan: Weaving the Tayal's Rainbow as an Example of the Tayal Tribe" (2007). Education courses, such as "Design and Establishment of Digital Teaching Materials for Traditional Tayal Art (Sub-Plan One): Research on Digital Narrative Teaching of Tayal High School Handicraft Cultural Creative Products" (2012).

The research above plan focuses on pattern analysis and visual image studies, which are the research directions concerned with the artistic characteristics of Taiwan's minority ethnic groups in recent years. The research institutions involved span various fields such as design, communication, business, leisure, and tourism. Among them, researchers from non-design fields are also involved, attempting to explore from the perspectives of business, leisure, and tourism how the patterns and images of Taiwan's minority ethnic groups are interpreted and then applied to the product development of the cultural industry. In the fields of design and communication, the research topics on patterns and images are, respectively, "Research on the Cultural Characteristics of Taiwan's Minority Ethnic Groups from the Color and Pattern of Textiles (I, II)" (2011, 2009) and "Research on the Visual Images of Taiwan's Minority Ethnic Groups: A Case Study of Paiwan and Rukai Myths, Legends and Rituals" (2011); "Research on Minority Ethnic Images: Photography, Memory and Local Perspective" (2007); "Research and Application of Taiwan's Minority Ethnic Image Culture in Product Design" (2004). Among them, the research theme led by Zeng Qixiong focuses on studying the colors and patterns of Taiwan's minority ethnic textiles. This is also the beginning of this research's participation in the primary survey. From this, the themes approved by the Taiwan Science and Technology Committee for research plans for studying the patterns and colors of minority ethnic textiles have received less attention, which undoubtedly opens another research path. In the early stage of this research, we had academic exchanges with Japanese institutions. We learned that Japan also has a long-term deep interest in the topic of Taiwan's minority ethnic groups. Therefore, we searched the CiNii-NII Academic Literature Database (a database of academic literature from the National Institute of Informatics in Japan)[4] for topics related to this research. According to the results of the database index, research on Taiwan's minority ethnic groups in Japan mainly explores the field of anthropology, including cultural anthropology, archaeological anthropology, physical anthropology, and linguistic anthropology. However, some research primarily focuses on the expression of ethnic art, such as Huang Guobin and Zeng He Yingzi, who analyze the pattern formation and color expression of minority ethnic textiles from the visual design perspective [49, 50]. Haruna Tanamoto and Junko Higashimura start from the weaving process, trying to understand the weaving industry of minority ethnic groups [51, 52]. Chizawa Riki and Tanaka Kiyoshi take traditional houses as research objects, exploring architectural styles and ancestral pillar patterns, respectively [53, 54].

What is worth noting is "Attempting to Classify the Diamond Patterns in the Traditional Textiles of Taiwan's Indigenous Peoples" (Trying to classify the diamond patterns in the traditional textiles of Taiwan's minority ethnic groups) [55]. And "The Evolution of Color Recognition in the Indigenous People of Southern Taiwan" (The Development Trend of Color Recognition in Minority Ethnic Groups in Southern Taiwan) [56]. These two papers analyze the pattern formation of traditional costumes of Taiwan's minority ethnic groups, focusing on the diamond patterns of different groups, using morphological similarity analysis to study the rules of pattern composition. The latter uses the Paiwan group as the research object, conducting surveys and analyses of commonly used colors, such as housing decoration, traditional costumes, color vocabulary, and other research topics. From this, some of the academic work

[4] CiNii-NII Paper Information Navigator Database (Academic Literature Database of the National Institute of Informatics, Japan), https://cir.nii.ac.jp, 23rd October 2022.

in Japan still has research content, trying to find materials related to design applications from primitive art. This can help us understand that other researchers are participating in this research direction in the design field, and their research results can also be used as a reference for this study.

2.2 Material Culture of the Pan-Atayal Group

2.2.1 Classification Overview

According to the research from a linguistic perspective by Li Rengui, the Atayal language group occupies the northern half of Taiwan, from Wulai Township in Taipei County to Ren'ai Township in Nantou County, where people of the Atayal language group live. Regarding language classification, the Atayal language group is first divided into Atayal and Seediq (Sediq) languages. The Atayal language can be divided into the Seqoleq and C'uli branches [57]. From a geographical distribution perspective, the Seqoleq dialect has the most expansive area, including Taipei, Yilan, Taoyuan, Hsinchu, Taichung, Nantou, and six other counties, which are also geographically continuous, followed by the C'uli dialect, which is mainly in the peripheral areas of Yilan, Hsinchu, Miaoli, Taichung, Nantou, and other five counties. The main dialects of the Seediq are all in Ren'ai Township, Nantou County, among which the Truku dialect (Truku) is spread throughout Hualien County. Lin Xiuche's research believes that the Atayal system has two sub-tribes: the Tayal sub-tribe (Tayal) and the Sediq sub-tribe (Sediq). Among them, the Sediq sub-tribe has three groups, including Tgdaya (the western branch in Nantou County is the Musha; the eastern branch in Hualien County is the Pulibau), Toda (the branch of the west in Nantou County is the Tuuda; the east branch in Hualien County is the Taosai group), Truku (the branch of the west in Nantou County is the Derlug group; the east branch in Hualien County is the Truku group). The ancestral homes of the three branches are all in Ren'ai Township, Nantou County. Three to four hundred years ago, each group had people migrating eastward, forming the current situation where the Sediq sub-tribe is distributed in both the east and west [58]. In addition, Liao Shouchen also mentioned the expansion routes of the three Atayal groups, namely the Segoleg system (Segoleg) moving north, the Tseoles moving northwest, and the Sedeg system (Sedeg) moving east [59].

In summary, Li Rengui and Lin Xiuche believe that the source of the Atayal language group is concentrated in Ren'ai Township, Nantou County, and began to differentiate and spread from this place. Liao Shouchen, based on the legend of the origin of the Atayal tribe, believes that in addition to the Tseole system originating at the junction of Hsinchu, Miaoli, and Taichung, the Seqoleq and the Seediq all originated in Ren'ai Township, Nantou County. As shown in Fig. 2.2, the range of Atayal migration can be understood as being vast, and the distribution of sub-tribes and tribes can be discerned from the migration paths of the sub-tribes [60]. At the same time, the migration process of the sub-tribes includes not only the movement of the tribespeople but also the migration and integration of material and culture. The place of origin and migration paths should be clues for observing cultural context. Another concern is that the era has gradually become distant. We can only indirectly understand the living conditions of the minority ethnic groups at the time from the field investigations of Japanese scholars from around 1895 to 1945. However, using the remaining cultural relics as research objects has become a pivotal clue to understanding the evolution of the lives of Taiwan's minority ethnic groups.

The Atayal and Saisiyat tribes in the northern mountains of Taiwan live adjacent to each other. Chen Qilu pointed out in his book "Ethnicity and Culture" that the Atayal and Saisiyat tribes may be the descendants of the earliest immigrants, who may have moved in during the pre-ceramic age, perhaps because their original habitat did not produce pottery, so they did not make pottery [61]. Feroli also mentioned that the Atayal tribe does not have pottery, does not believe that their ancestors ever made pottery, and does not have decorative wood carvings and stone carvings [62]. From this, it can be understood that the decorative patterns of traditional Atayal costumes are the most important cultural symbols of the tribe. Although the Saisiyat

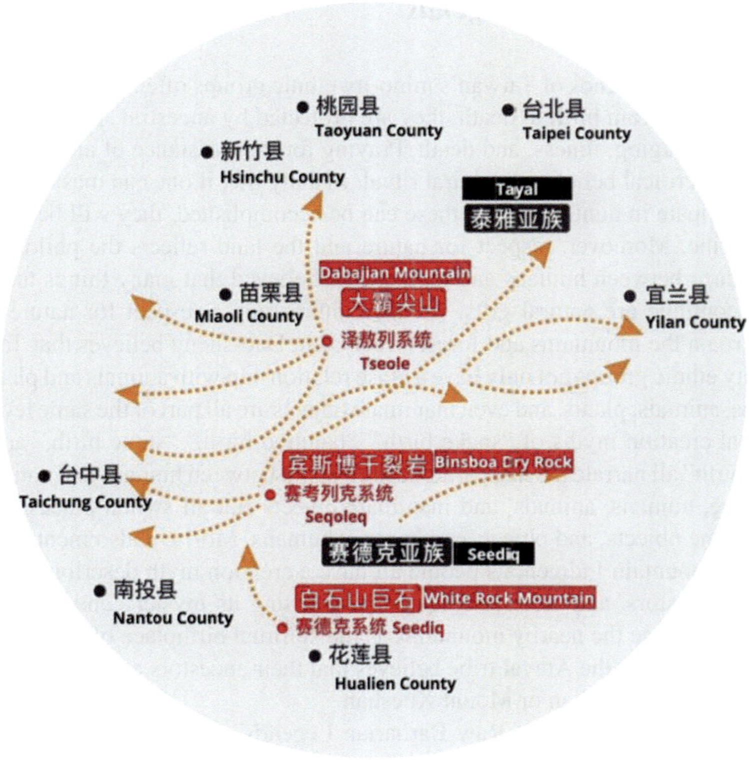

Fig. 2.2 Schematic of the origin and migration paths of the Atayal and Seediq (drawn by this study)

tribe, which also belongs to the Paiwan language group, has adopted the Atayal type of tattoo, the Atayal tribe is the only one among Taiwan's minority ethnic groups with the custom of facial tattoos. Daxiu Wulawan Bima mentioned that traditional Atayal men tattoo forehead patterns (forehead) and chin patterns (chin), and women tattoo forehead patterns and cheek patterns (cheeks). The forehead patterns of both men and women are continuous in horizontal parallel lines, one to three for men and three to five for women on the forehead. Each band pattern is divided into several sections; only one is tattooed in the middle of the man's lower chin. Women have no chin patterns, and their cheek patterns start from the root of the ear and extend to the middle of the lower chin, with parallel lines on the edges and cross diagonal lines in between to form a net. All facial decorations' shapes, sizes, features, etc., often vary depending on the sub-tribe and group [63]. According to the survey data of "Taiwan's Aboriginal Research," professional tattooists, if they make mistakes in the tattoo style or cause facial infections, will be subject to sanctions to stop tattooing and be expelled [64]. Not only is facial tattooing a symbol of the Atayal tribe, but those who professionally administer these tattoos cannot afford any mistakes.

2.2.2 *Mythology and Legends*

The myths and legends of Taiwan's minority ethnic groups often revolve around the theme of life. From birth to death, they are protected by ancestral spirits, especially during birth, aging, illness, and death. Praying for the assistance of ancestral spirits becomes a critical belief and cultural ritual. In daily life, if one can master life skills and participate in hunting, and if these can be accomplished, they will be respected by the tribe. Moreover, respect for nature and the land reflects the philosophy of coexistence between humans and nature. It is believed that many things that allow life to continue are natural gifts. As such, cultivating a respect for nature allows one to roam the mountains and forests freely. Lin Dao-sheng believes that Taiwan's minority ethnic groups not only have a close relationship with animals and plants, but humans, animals, plants, and even inanimate objects are all part of the same level. The ancestral creation myths of "snake birth," "bamboo birth," "stone birth," and even "dung birth" all narrate the inseparable relationship between humans and nature [65]. Therefore, humans, animals, and inanimate objects can all switch places; humans can become objects, and objects can become humans. Mori Uji also mentioned that Taiwan's mountain Indigenous people all have a creation myth describing the birth of their ancestors, and there is a custom of praising its mystery and honor. Their legends often use the nearby mountains as the spiritual birthplace of their ancestors [66]. For example, the Atayal tribe believes that their ancestors came from the giant rocks of Mount Dabajian or Mount Xueshan.

The book "Collection of Raw Barbarian Legends" points out the racial origin legend of the Musha Seediq tribe, whose ancestors were born from a divine tree that was half wood and half stone. One day, a man and a woman came out of the tree, and their union produced offspring [67]. Liao Shou-chen pointed out that there are three birthplaces in the Atayal tribe's legends, namely Mount Dabajian (Papa-waga) (the junction of Hsinchu County, Miaoli County, and Taichung County / Jianshi Township, Hsinchu County), Pinsbukan Split Rock (the upper source of Beigang River / Ren'ai Township, Nantou County), and Bunobon Giant Rock of Baishishan (southeast of Musha / Ren'ai Township, Nantou County). Among them is the Bunobon Giant Rock of Baishishan; it is said that a large tree clung to this giant rock in the past, and the tribe was born from the rock burst by the tree roots. This racial origin legend is like the Musha above tribe's mixed birth of tree and stone [68]. In the book "Atayal Mythology and Legends," Dasiwulawan Pima mentioned that the Atayal and Seediq tribes were born from the "Great Tree," believing that the Great Tree symbolizes ancestral spirits. After death, they will return to the Great Tree and be protected by ancestral spirits [69].

For instance, in the 1930 Musha Incident, the Seediq tribe, unwilling to accept defeat and surrender, chose to commit suicide in the trees. This intense behavior is also related to their traditional belief in protecting ancestral spirits. According to the oral account of Jian Yunsheng, the Atayal tribe also has a well-known myth of the Rainbow Bridge. Men used to be headhunters, women were good at weaving or housework, and after death, they could pass the Rainbow Bridge without hindrance. The colorful horizontal stripes can be understood from their clothing as a reference to this legend. The horizontal stripes symbolize the Rainbow Bridge leading to the ancestral land. This concept has shaped the cultural characteristics of the Atayal tribe, with men defending their homes and women excelling in housework.[5]

Between 1895 and 1945, the Temporary Old Custom Investigation Committee of the Taiwan Governor's Office investigated the facial tattooing legends of the Atayal tribe. For the Maliguwan tribe, the origin of women's facial tattoos was that once, many women died at the same time, causing great grief to their ancestors. One night, a spirit appeared in a dream and told the ancestors that those with facial tattoos could avoid disaster. Since then, women have adopted the custom of facial tattooing. For the origin of men's facial tattoos, in the past, the ancestors divided into two teams, and the community chose which team to join, then determined the number of people by shouting. As a result, one side had more people, and the other had fewer. At that time, the team with more people said they knew how to make knife blades, so they needed more people. After saying this, they left the mountains and went to the plains. The team with fewer people, because of their small number, believed that they could not kill each other; otherwise, they would be extinct, so they tattooed their faces for identification. For the Beishi tribe, a man said to an ugly woman, "Your face is so ugly; maybe it will look better if you tattoo it." So, the woman agreed to the man's suggestion. The man first collected some torch charcoal and, at the same time, drew patterns on the woman's clothes, teaching her how to tattoo her face. The woman

[5] Jian Yunsheng, born on June 7, 1929, from the Atayal Tseole Wen water group, originally from Daxing Village, Tai'an Township, Miaoli County, once served as the head of the Rongan tribe. In May 1981, he established the Atayal Traditional Art Studio (Private Museum) at his home, where he began to collect minority artifacts and replicate bamboo and rattan crafts, i.e., traditional hand tools used by the older generation. He published the book "Atayal Traditional Artifacts" in 2007 and has long been dedicated to researching and preserving ethnic cultural history.

asked the Han people for a thin iron wire to prick her face, then smeared the torch charcoal on her face as decoration. In the past, women died quickly, but since they started tattooing, the death rate has significantly decreased. When the ancestors were deceived by the Han people and lost many people, they also thought that if men also tattooed their faces, they could clearly distinguish between friend and foe, and there would be no fratricide, and they might live as long as women. Besides, it wouldn't look good if only women had tattoos and men didn't, so men also adopted the custom of facial tattooing [70]. The legend of facial tattooing is also circulated in different stories among other subgroups of the Atayal tribe. Still, facial tattooing was initially for aesthetic needs, a sign of adulthood, and tribal identification. Later, the meaning of facial tattooing in the Atayal tribe has shifted to a symbol of self-assessment of abilities, such as men needing to have the experience of headhunting before they can tattoo their faces and women needing to have the ability to weave before they can tattoo their faces.

2.2.3 Clothing Culture

Culture is the way of life passed down by a nation, accumulating the country's unique style. Li Yiyuan believes that observable culture includes material culture or technical culture, such as using will and strength to conquer nature, obtaining the materials necessary for survival, using social groups to regulate and establish corresponding laws and regulations, or even overcoming fear through the spiritual level to produce cultural symbols including art, music, performance, etc. [71]. According to the "cultural model" proposed by Li Yiyuan, the thinking process of primitive human development has universal common characteristics. In the religious belief center of supernatural things of the Daxi Wulawan・Bimati and the Atayal tribe, the concept of the dead spirit tux (gods, demons) is essential, and the idea of fate tsinun tux (spirit woven) is also valued. The ancient Atayal's tux also carried the meaning of the dead and the dead spirit. Later, the concept of the living spirit was developed, also known as tux; the presiding spirit supports the body, and after leaving it, it becomes a dead spirit [72]. In Atayal's Ga-Ga (social norms/group organization), tux is believed to be afraid of red, so the tribe members wear dark, red-dyed clothes on their bodies, which scares ghosts [73].[6] On the pattern before the clothing, a diamond pattern is woven to symbolize the eyes of the ancestors, which has a protective effect. The pattern on the back of the clothing determines the tux with complex woven patterns.

The patterns of general natural tribes all follow the forms left by their ancestors, passed down from generation to generation, and imitated. Due to the plagiarism of one generation after another, the result will inevitably become distorted and another variant form [74]. Although it is difficult to determine what the common simple

[6] Ga-Ga refers to the general term for all norms, as well as the general term for customs and habits. Cited from Hei Dai・Bayan: "Exploring the Lifestyle of the Atayal People: An Atayal's Personal Testimony," Hsinchu County Cultural Affairs Bureau, Taiwan, 2002 edition, page 36.

geometric patterns come from, they are not something cut of nothing, and they are probably due to the inspiration of nature or the influence of technology [75]. Ni Jianlin believes that abstract geometric decorative patterns ' significance lies in that they can be a symbol of collective identity and even express connotations. Still, this symbol is not obtained through the association of the image itself but through the recognition within the group [76]. The cultural connotation of Atayal clothing comes from understanding traditional beliefs. Through imagination, colors and patterns are transformed into lines or shapes that do not represent any object images. These non-figurative geometric patterns become part of the group spirit and protect the tribe members. Therefore, geometric patterns are often seen in the decorative patterns of traditional Atayal clothing.

2.2.4 Weaving Technology

Like other minority ethnic groups in Taiwan, the Atayal people primarily use ramie (nettle) as the fiber for their textiles. Making ramie thread, from planting to harvesting, takes at least three months. After harvesting, the ramie is scraped, sun-dried, cleaned, and dried again before it can be twisted into thread, accumulated, sorted, boiled, rinsed, and sun-dried to become a usable thread. In the past, the pigments used for dyeing were derived from plant roots, extracting their juice and boiling it with the thread. For example, sweet potato juice can dye the thread a tea-brown color, while the black thread is achieved by adding ash juice, mud, or crushed crape myrtle leaves. Ochre is achieved by adding ash juice or by mashing the tuber of the cassava and adding water. In the book "Survey Report on the Customs of the Aboriginal Tribes: Volume One, The Atayal," it is mentioned that the Atayal did not have red, blue, yellow, black, and other dyes. These colored cloths were all obtained through the exchange or woven with exchanged wool (sometimes, the colored threads from the exchanged cloth were unraveled and used) [77]. Today's minority ethnic groups in Taiwan have replaced ramie thread with cotton thread, using cotton for the warp and weft to form the thread. The weft is usually made of wool, so the woven part that forms the colored pattern has a thick texture [78].

The weaving tools of Taiwan's minority ethnic groups use a "horizontal backstrap loom," with women responsible for the weaving. In the early days, the Atayal people of the northern mountains of Taiwan were skilled at using the horizontal backstrap loom to weave cloth. The structure of this loom is simple, but it can produce complex patterns. Chen also pointed out that the horizontal backstrap loom is found in Egypt, Sudan, and the Tanzanian territories of East Africa, from Mexico to the Andes in America and Central Asia to Southeast Asia, with the densest distribution in Southeast Asia [79]. The Atayal people also have another weaving tool called a "bow loom." First, a bamboo strip is split, modified, roasted, and bent into a bow shape, using the bow's tension to create a temporary weaving machine. The bow loom can weave a band about 50 cm long and 5 cm wide. The used bow can be discarded, and a new one can be made from local materials. Atayal men usually use bow-woven products to make the straps for hunting backpacks or tying their machetes.

Traditional Atayal textiles rarely feature figurative or realistic animal and plant patterns, mainly composed of diamond patterns and their variations or other geometric designs. Common patterns are arranged left and right to form compound stripes, and the complex pattern formation is also a characteristic of Atayal textiles. Traditional Atayal weaving techniques, such as plain weave, the foundation of weaving, and the weave structure formed by the warp and weft alternately sinking and floating, are shown in Figs. 2.3 and 2.5. Twill weave, the intersection of warp and weft on the fabric's surface, presents a twill structure. Refer to Figs. 2.4 and 2.6. Float weave: according to the weave, the weft is exposed outside the warp. Picking, according to the pattern design, calculates the number of warp yarns and uses the picking hook or small shuttle to sequentially pick up a part of the warp to form a pattern. Jacquard, with a large pattern organization, the resulting fabric pattern is more significant, and the pattern is more complex, such as small diamond patterns organized into complex large diamond patterns. In the book "Taiwan's Minority Ethnic Textiles: Weave Structure and Pattern Analysis," it is mentioned that the main weave structure types of Atayal textiles are plain weave, changing twill, combination organization, multi-color weft show flower organization, warp float organization, particular warp organization, unique weft organization. Although there are many branches of the Atayal tribe, the pattern features are based on diamond patterns and stripes as essential elements for combination and change [80]. Okamura Yoshiemon and Zhang Yongxin pointed out in the book "Taiwan's Aboriginal Cloth" that the organization pattern of Atayal textiles is the most varied among Taiwan's minority ethnic groups, and the Saisiyat tribe and the Atayal tribe are slightly similar [81].

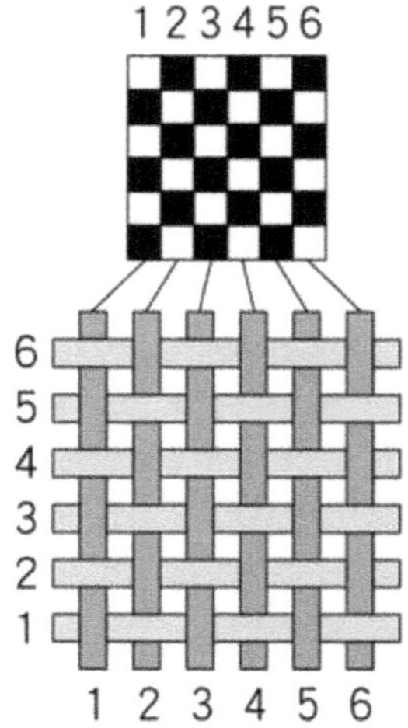

Fig. 2.3 Plain weave organization (drawn by this study)

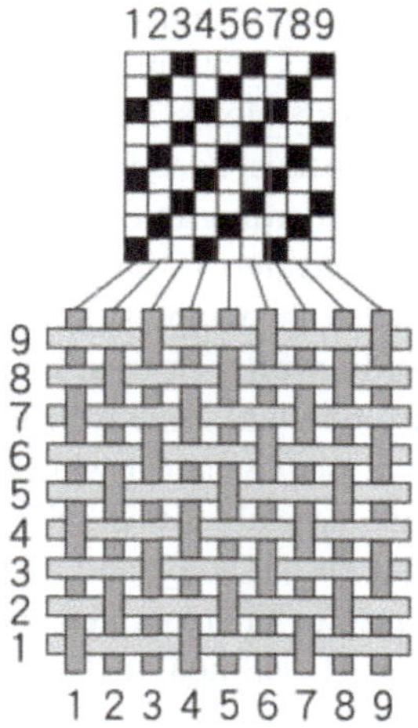

Fig. 2.4 Twill weave organization (drawn by this study)

Fig. 2.5 Plain weave (part of a woman's long-sleeved short top) (photographed by this study)

Fig. 2.6 Diamond twill weave (part of a woman's foot protection cloth) (photographed by this study)

2.2.5 Traditional Costumes

Traditional Atayal clothing is a combination of sewn and draped forms. Traditional clothing can be divided into everyday wear, work, and ceremonial attire. There is no obvious class distinction in traditional Atayal society. The attire of the headman's family is the same as that of the tribespeople. The essential accessories of men's clothing include headbands, headdresses, breastplates, tops, capes, front modesty cloths, and knife belts. The critical accessories in women's clothing include headbands, headdresses, breastplates, tops, capes, skirts, and leg wraps.

According to Suzuki's research, the clothing of the Atayal tribe can be roughly divided into seven categories, including tube sleeves, vests, square cloths, breastplates, leg wraps, skirt cloths, and cloaks. Men's and women's clothing are cut differently, with men wearing hats and women wearing headscarves. The hats are woven from rattan, some have brims, and some are decorated with bear skin (the part with a crescent-shaped white hair on the neck) or pottery buttons [82]. Lisa mentioned two types of Atayal shawls: a five-foot three-piece cloth, like a cassock, wrapped flat on the body, and a three-foot two-piece cloth, slung diagonally [83]. Hu Jiayu and Cui Yilan pointed out that the long dress is one of the most basic clothing styles for Atayal men, including sleeveless and sleeved. The short dress only reaches the waist and is exclusively for men to wear when going out. It is sometimes worn over a long dress. Waist skirts are paired with sleeved short dresses, and men generally do not wear waist skirts. Both men and women can wear tube sleeves, breastplates, and cloaks, but due to gender differences, the way they are worn is also different [84]. The survey reports related to the period from 1895 to 1945 also mentioned that the Atayal tribe's dressing method has a mix-and-match situation. Its costumes have "lupus" (long dress) and "rating" (short dress), and two types of differences lie in the material, length, pattern, color, decoration, etc., so they have different names. If it is not a man who has hunted the enemy's head and a woman who is not good at weaving this garment, it is not allowed to wear a short dress on the long dress, but it is okay to wear it underneath [85].

Except for the Yami tribe (Tao tribe), all other tribes in Taiwan have used shell bead clothes (lukus kaxa) (giant clam), especially the Atayal, Saisiyat, and Paiwan tribes use them more widely. The shell bead clothes are adorned on the cloth with two functions: one is as a trading medium, that is, it is used as currency in society, and the second is as a decoration on the clothes. The shell beads used as a currency function are only seen in the Atayal tribe [86]. Dudbridge also mentioned that long ago, the minority tribes in Taiwan had a peculiar currency. The currency was shaped like a flat disc, the inside was composed of regular spiral lines to form a curling shape, and the shape and size represented different currency values. These currencies are mainly distributed in the tribes of the coastal areas [87]. Therefore, it is speculated that this kind of shell currency should refer to the decorative items commonly seen in the Paiwan or Rukai tribes in southern Taiwan. Only the traditional costumes of the chiefs or nobles can be used.

2.2.6 Clothing Shape and Geographical Distribution

The Atayal tribe's concept of color can be found in the vocabulary describing colors, showing the similarities and differences among various subgroups, such as the

Galaide tribe, Shegaru tribe, Bunsui, Dahuhu tribe, Taranan tribe, and Qinaer tribe. The color vocabulary includes black, white, red, yellow, green, blue, and purple [88]. The Musha, Taotuo, and Zhuoluo tribes have color vocabulary, including black, white, red, yellow, green, blue, and purple. The Inner Truku, Outer Truku, Taosai, Bizhuolan, and Pulibao have color vocabulary, including black, white, red, pink, yellow, blue, and purple [89]. The Msbtunux and Dgiyaq tribes have color vocabulary, including black, white, red, yellow, green, blue, and purple. The Maligu Bay and Hakul (Hakku) have color vocabulary, including black, white, red, yellow, and blue. The Beishi has color vocabulary, including black, white, red, yellow, blue, and light blue. The Nanshi has a color vocabulary, including black, white, red, yellow, green, blue, and brown. The Sijia Yewu, Plnawan, and Klesan have color vocabulary, including black, white, red, yellow, green, and blue. The Meiyuan tribe has a color vocabulary: black, white, red, yellow, blue, and purple. The Mnibu (Sikikun and Piyahan communities) have color vocabulary, including black, white, red, peach, yellow, green, blue, and purple [90]. Modern scholars like Li Yiyuan have also surveyed Klesan's descriptions of color knowledge, finding color vocabulary including white, light yellow, grey (all the same word), black, dark blue (all the same word), red, and brown. If they encounter other colors or need to determine a close color, they describe it in terms of various things, such as describing light blue as the color of the sky [91]. From this, it can be inferred that the Atayal tribe has long used a variety of colors in daily life, with the color vocabulary of black, white, red, yellow, green, blue, and purple being more numerous. Through these color vocabularies, we can also explore the Atayal tribe's color usage habits. The book "The Tribes and Migration of the Austronesian Peoples of Taiwan" also points out that in the early Austronesian period, it seems that only "black" and "white" were distinguished. Black and dark are related. White and clean often come from the same linguistic root, but the Austronesian languages of Taiwan distinguish black, white, red, yellow, and blue and usually do not distinguish between blue and green [92].

The patterns and colors of the Atayal tribe's square clothing collections, black series (Klesan in Yilan), white series (Wulai in Taipei, Jianshi in Hsinchu, and Hualien area), red series (Musha in Nantou), all three-color series have wave patterns and straight-line patterns. In the large decorative bands, the former's large patterns, such as the "belt-shaped olive pattern" and "triangular wave pattern," have a linear flow performance, the raised pattern interwoven with the tu-iuk (pattern name), the revealed shadow pattern is also composed of simple thin lines, linearity is the main style of this color series pattern. Compared to the latter's large patterns, such as the "compound diamond pattern" and "quadruple diamond pattern," the latter's pattern style is based on geometric diamonds, with diamonds as the building units, presenting a planar style [93].

According to the records in "*Sino-Japanese Weaving Craft Exchange Exhibition Taiwan Pavilion*" [94], "*Taiwanese Minority Ethnic Clothing Culture: Tradition Meaning Illustration*" [95], and "*Taiwanese Minority Ethnic Body Decoration and Clothing*" [96], the Atayal tribe, due to its wide distribution, has led to differences in patterns and colors of traditional clothing. For example, the Tayal Msbtunux (Fuxing Township) often uses black and dark blue threads to decorate ceremonial clothing.

The diamond patterns of their weaving are single-colored and detailed, with large, varied units of combination. The Klesan (Klesan Township) prefers red and black but also likes to pair sky blue with peach red. During major ceremonies, both men and women wear beaded clothing. The Truku of the Seediq (Xiulin Township) mainly uses plain white. Men wear sleeveless long shirts and sleeve covers; women wear sleeveless long plain dresses, add sleeve covers, wrap the skirt on the left and right, and then tie it with a belt; the skirt and shawl are woven with black, red, peach red, and cyan, with scattered small diamond patterns.

The geographical distribution of the various sub-groups of the Atayal tribe and the characteristics of their clothing are undoubtedly important clues to differentiate the differences between the sub-groups. However, previous research surveys have rarely compared the similarities and differences in the patterns and color matching of the sub-group textiles. Moreover, the shape and structure of the patterns are indispensable decorative elements in the traditional clothing of minority ethnic groups. Therefore, a comprehensive discussion of the characteristics of the patterns and the evolution of their shapes and systems has become one of the focuses of this study.

2.3 Visual Form Composition and Combination

2.3.1 Origin of Decorative Patterns

From the visual design perspective, we can understand the knowledge system of decorative patterns and pattern structures through the language of form. Then, we can explore the composition of Atayal textile patterns using the design thinking of form modules. Using visual design to explore the unique weaving expressions of the Atayal tribe can also provide new topics for design education.

The course of human civilization development is the emergence of material culture. The transition from manufacturing to the production of materials is practical and rich in decoration, using different patterns, colors, and shapes to reflect their aesthetics and consciousness. The development of arts and crafts has a distinct ethnicity, and the arts and crafts culture of each ethnic group reflects the ethnic group's pursuit of beauty, including the creation of decorative patterns. In the book "The Composition of Beauty" (The Composition of Beauty), it is pointed out that relics and ancient tombs were discovered worldwide, and the unearthed artifacts can all be seen as simple geometric pattern decorations. The earliest decorative patterns used on utensils and daily necessities are called practical goods decorations; those used on the human body are called tattoo decorations; those used in the dwelling place are called the most primitive environmental decorations [97]. Zuo Han mentioned that the geometric patterns of abstract shapes originated from the transplantation of the structure of woven fabrics. As woven fabrics were the supports for pottery clay during pottery making, primitive people gained the experiential form of geometric patterns from the imprints of woven fabrics, promoting the ability to summarize abstract authentic images with simple lines [98]. According to the definition of "decoration" in the "Chinese Arts and Crafts Dictionary," it generally refers to adornment or dressing up. "The Book of Later Han - Biography of Liang Hong" states that the woman (Meng Guang) asked to make cloth clothes, straw sandals, weave baskets, and spinning tools. When she married, she started decorating and entering the door [99]. From the literature above, it can be understood that decoration has become a style of art, especially in the art category; decoration itself can be regarded as a pattern and is applied to the surface of objects. Zhang Tianxing and others mentioned that decorative patterns refer to patterns or graphics with decorative properties. It is a creative art design form with craft materials and production technology. Decorative patterns can be divided into figurative, abstract, and composite forms [100].

Primitive patterns are not just decorative patterns. The most important thing is that they also have symbolic meanings. In this sense, primitive patterns are also a language for conveying messages or communication [101]. Stone also gave an example of Arabic numbers and dates sometimes woven into carpets in Arabic calligraphy [102]. From this, text can also be a subject of decorative expression.

In contrast, the languages of Taiwan's minority ethnic groups do not have written characters but have similar symbolic expressions. In terms of the ornamental patterns of Taiwan's minority ethnic group's woven products, Ba Yici[7] believes that the traditional clothing patterns of the Atayal tribe (Msbtunux) in Fuxing Township, Taoyuan County, contain textual messages, as shown in Table 2.1 [103]. Whether these patterns, as Ba Yici said, represent the traditional characters of the Atayal tribe, there is currently no further research and can only be explained speculatively.

The patterns of Atayal woven products are mostly diamond-shaped variation patterns. The tribes generally interpret these patterns as the eyes of ancestral spirits. Relevant literature points out that different families and places have their exclusive pattern styles [104]; specific patterns can only be used and inherited by the original weaver unless they are obtained through the ceremony of buying technology by exchanging money or goods. Otherwise, they cannot be imitated. The book "Survey Report on Aboriginal Customs: Volume 1 Atayal" also mentions that the women of Taranan, Mnibu, Msbtunux, and Gogan protect the design patterns of women's woven products. Those who imitate their patterns without prior notice to the pattern designer are subject to a ransom of a sickle or a hoe [105].

Records related to exclusive textile patterns have been lost, and it is impossible to confirm the family or author to which the pattern belongs. We can only speculate about the region where the pattern may have existed.

---

[7] Fr. Alberto Papa (Bai Yici), born in 1935, is an Italian Franciscan priest who has long served in the mountainous areas of Taiwan. In 1965, he arrived in the mountainous area of Fuxing Township, Taoyuan County, to preach and served at the Sanmin Catholic Church in Sanmin Village, Fuxing Township. Due to the language barrier, he worked hard to learn the Atayal language and once translated the Bible into Atayal using Roman phonetics. He discovered that the symbols in the traditional weaving of the Atayal tribe, known as BIRU, might be the lost Atayal script, and based on this, he compiled Atayal language textbooks. His works include "Atayal Grammar" (1986), "Atayal Language Study Beginner's Book Volume One" (1995), and "Atayal Language Study Beginner's Book Volume Two" (2000).

Table 2.1 Ba Yici's interpretation of the implied meaning of Atayal tribal textile patterns in Fuxing Township, Taoyuan County

Textile Pattern	No	Symbolic Marriage Implication	Pronunciation
	1	Male	qutux
	2	Female	sajin
	3	Describes the upward and outward development of life (Parents)	tugal, ciugal
	3	Describes the inward and deep development of life (Children)	tugal, ciugal
	4	Male (Without informing parents, the relationship is not established)	pajat
	5	Female (Informing parents, relationship established, and marriage can be discussed)	magal
	6	Parents of both sides communicate	tjiu, cjiu
	7	Parents' blessing	pitu
	8	Announce engagement	spat
	9	Marriage, establishing kinship	qeru
	10	Harmonious relationship	mpu

2.3.2 *Pattern Organizational Structure*

The pattern composed of weaving structures is also a kind of rational thinking expression directly related to shaping, and the weaving pattern composition has similarities with design elements. Stone mentioned the relationship between the shape and structure of the weave pattern. The weave pattern comprises the warp and weft's vertical and horizontal intersecting grid lines. The square represents the same density of warp and weft. Taking the circle in Fig. 2.7ⓐ as an example, the more grids the circle uses, the more warp and weft lines form the circle. Compared with the circle weave pattern with fewer grids, it will inevitably present a more detailed circumference line. Otherwise, the fewer the lines, the more jagged appearance, or the interior will present a cross-shaped coarse circle, as shown in Fig. 2.7ⓑⓒⓓ [106]. The weave pattern's shape comes from the grid density arrangement. Points, lines, surfaces, bodies, textures, and colors are the elements of design, which are the simple

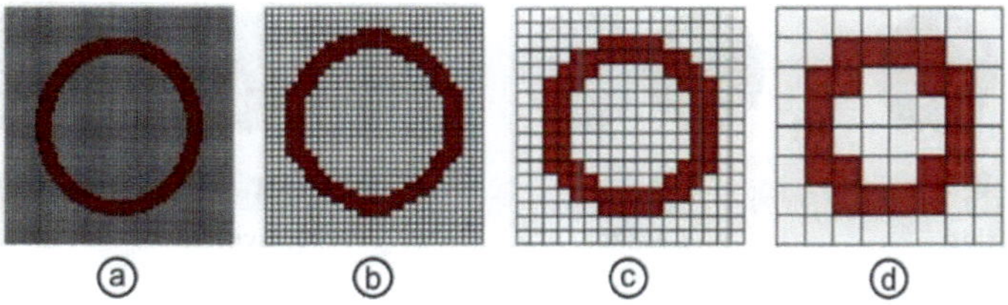

Fig. 2.7 Relationship between weave pattern shape and structure (drawn by this study)

components after decomposing the components that make up the design [107]. Lin Chonghong mentioned the method of plane graphic composition, which is like the principle of aesthetic form, but the plane graphic is expressed through visual perception. Visual acceptance will produce exceptional images due to people's psychology because "shape" is the most basic object in visual phenomena. For example, repetitive shaping comprises regular, repetitive units or irregular combinations, expressing a unified visual perception [108]. Yang Qingtian believes that the principle of shaping composition is that "shape" is the element that makes up the form, and "form" is the style of composition. Although the two are two sides of the same body, in terms of the elements of form, exploring the nature of "shape" is still the priority [109].

Wang Wuxie and Liang Juting further explained that repetitive images can be called basic shapes. Design conception often starts with a specific image as the basis, then uses repetition or other methods for appropriate expression, hence the term "basic shape." The basic shape does not necessarily have to be repeated. It can also be approximated and gradually changed [110]. The combination of form order that constitutes the form, such as the arrangement method of the basic shape, can be divided into absolute repetition (basic shape repeated arrangement), interval repetition (positive and negative alternation of repeated, basic shapes), rotation repetition (repeated, basic shapes change position in direction), reflection repetition (repeated, basic shapes are mirrored up and down or left and right), as shown in Fig. 2.8 [111]. There are also eight other ways of expressing the relationship between images, such as separation, contact, overlap, transparency, union, reduction, difference, and nesting. Each type of encounter leads to different spatial effects, as shown in Fig. 2.9 [112].

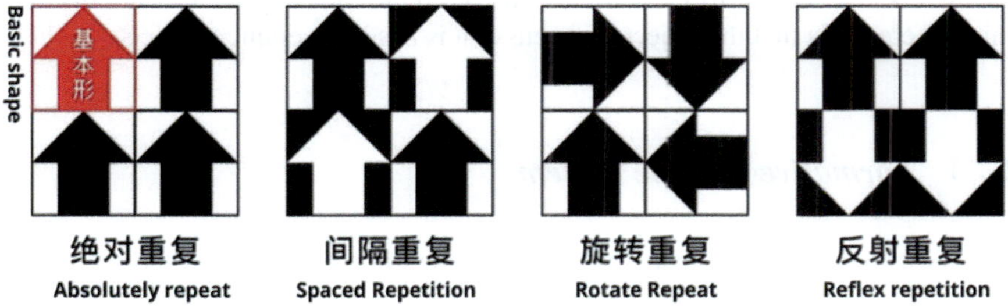

Fig. 2.8 Morphological order combination method (drawn by this study)

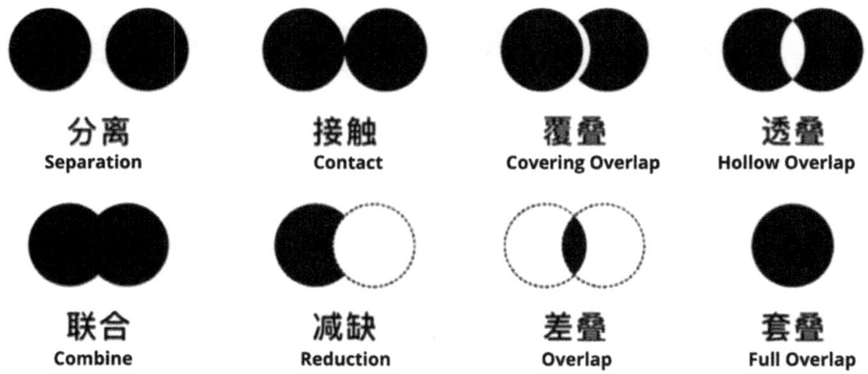

Fig. 2.9 Eight ways of image encounter (drawn by this study)

Wax Raiser and Buschick Snyder also exemplify the method of "creating new formative structures," where the original forms of graphics should have similarity and identity so that they can be freely connected, such as graphics touching each other, overlapping, or dividing to create new graphics [113]. The organization of patterns solves the problem of pattern composition, emphasizing the relationship between the whole and the parts and the arrangement and layout to express the theme as perfectly as possible. As a collection of multiple complete images, the pattern manifests explicitly in resolving the relationships between pictures regarding dominance, density, reality, height, and correspondence and subsequently adjusting the images' size, shape, length, curvature, and dynamics [114]. From this, pattern composition embodies regularity and order, generating a strong sense of rhythm and rhyme between images.

The composition layout of textile patterns, also a manifestation of pattern organization structure, can be analyzed through understanding the knowledge of formative design. For example, the patterns of ethnic minority textiles often feature repetitive patterns. This repetition is necessary for the pattern system to present as orderly and complex. In the patterns of ethnic minority textiles, progressive repetition is built on a relatively simple organizational structure, thus forming a complex structure. As mentioned above, design elements, repetitive formation, basic shapes, and image encounters are like the formation methods of ethnic minority textile patterns. The only difference is that the object of discussion is based on ethnic artifacts.

2.3.3 Formative Module System

Modularity, or module-based, is a structural rule often appearing in more complex morphological systems. Such characteristics can be observed if the whole is deconstructed into several small unit objects. Although hiding the internal subtle structure, the overall performance of modular design can still be observed to be based on simple shapes combined with other module units to form a larger whole [115]. Although Taiwan's ethnic minorities do not know about modular design, from the perspective of the formative structure of textile patterns, it is familiar to produce many modular pattern formations through the norms of weaving or embroidery techniques and the rules and procedures of application techniques. Analyzing the formal structure of Taiwan's ethnic minority textile patterns, we see that these patterns have their structural combination features. If the patterns are further deconstructed, they can be restored to the original basic units, and it may be possible to discover the combination rules of the formation of woven patterns, thereby exploring the structural relationship of the patterns.

When focusing on pattern features, another element that constitutes textiles, its color, is also one of the keys to the rich expression of patterns. At this point, the factors of color must be excluded first to analyze the formation of patterns based on the principle of uniformity. Gombrich also mentioned that the analysis of patterns should minimize color interference as much as possible. By eliminating complex factors, the simple forms existing in the pattern can be obtained, such as using only one color to emphasize the order characteristics of its structure, especially single-color patterns, which are more accessible to recognize than multi-color patterns [116].

The study of form in the past has often focused more on morphological psychology. In contrast, the organization or management of form received less attention until the development of technology pointed to the digital age, that is, the sign era. The development trend of "Future Studies of Form" has gradually appeared in a systematic change of form, so there has been a different discussion on the knowledge of form from the past. Changing the systematic unit body [117], with a fundamental geometric shape, gradually produces new form changes through different levels of form superposition, which can create many different form styles. Still, these forms have similar or identical unit-body relationships. The book "Form Morphology" also mentions that the method of form creation can be carried out through modularization [118], the five stages of systematic form evolution, from top to bottom, each stage of change is a process from simple to complex, thus creating a series of systematic forms, each form implies common vital elements. Knight pointed out in his research "Transformations of Languages of Designs" that grammar is shaped in form transformation, at least by two separate components, according to specific operation rules, to complete a new combination. Hence, the new combination becomes a new grammar, which can be deduced to derive new shapes continuously [119].

One must understand the detailed structure of "shape" to analyze modular form. For example, the patterns of Taiwan's minority ethnic textiles exist in the self-similarity of shape, as if hiding a kind of orderly rule. These patterns based on form orders are composed of similar parts of the whole, which can be extended due to pattern operation. Taking the literature above on modular form as an example, from the analysis of individual shapes in the whole form system, it can be found that "shapes" coexist with similar basic shapes, which can be understood as the essential elements that constitute the form system. The expression of form design often repeats the same form or continues in all directions. However, the pattern of Taiwan's minority ethnic textiles and form expression is different. The learning process of minority ethnic weaving technology results from generations of weaving technology learning, which differs from form design learning. Still, there are similar parts between the two. Therefore, through the design thinking of systematic form, not only can the design context of minority ethnic textile patterns be analyzed, but it can also help the learning and application of design basics.

References

1. Suzuki Zhi: "*Taiwanese Minority Customs (台湾少数民族风俗),*" translated by Wang Meijing, Taiwan Indigenous Culture Publishing House, 1999 edition, page 25.
2. Japan Shunyi Taiwan Minority Research Association: "*Overview of Taiwan Minority Research (台湾少数民族研究概览),*" Japan Wind Echo Co., Ltd., 2002 edition, page 10.
3. Xiao Qiongrui: "*Islanders・Customs・Paintings: 18th Century Minority Life Images (岛民・风俗・画: 18 世纪少数民族生活图像),*" Taiwan Dongda Book Company, 1999 edition, page 148.
4. Howard Morphy, "*Aboriginal Art,*" London: Phaidon Press, 1998, P.14.
5. Wang Songshan: "*Contemporary Taiwan Minority Art (当代台湾少数民族艺术),*" Taiwan Art Education Museum, 2001 edition, page 11.
6. Jiang Xun: "*Meditations on Beauty: Preliminary Discussions on Chinese Art Thought (美的沉思:中国艺术思想刍论),*" Taiwan Lion Book Co., Ltd., 2003 edition, page 7.
7. Xu Gongming: "*Minority Art and Museum Display (少数民族艺术与博物馆展示),*" Taiwan Nantian Bookstore Co., Ltd., 2004 edition, page 48.
8. Suzuki Zhi: "*Taiwanese Minority Customs (台湾少数民族风俗),*" translated by Wang Meijing, Taiwan Indigenous Culture Publishing House, 1999 edition, page 25.
9. Daxi Wulawan・Bima: "*Taiwan Indigenous Culture Series: Taiwan's Minorities Amis, Atayal, Saisiyat; Puyuma, Shao, Paiwan, Tao, Rukai; Bunun, Tsou (台原文化丛刊系列:台湾的少数民族阿美族、泰雅族、赛夏族;卑南族、邵族、排湾族、达悟族、鲁凯族;布农族、邹族),*" Taiwan Indigenous Publishing House, 2001, 2002, 2003 editions.
10. Lin Jiancheng: "*Taiwan Minority Art Field Notes (台湾少数民族艺术田野笔记),*" Taiwan Artist Publishing House, 2002 edition.
11. Chen Yulan: "*Taiwan's Minorities (台湾的少数民族),*" Taiwan Hiking Culture Co., Ltd., 2004 edition.
12. Wang Songshan: "*The Society and Culture of Taiwan's Minorities (台湾少数民族的社会与文化),*" published by Taiwan United Publishing Co., Ltd., 2001 edition, page 48.
13. Wang Hailong: "*Visual Anthropology (视觉人类学),*" published by Shanghai Literature and Art Publishing House, 2007 edition, page 62.
14. Bernice Martin: "*Contemporary Society and Cultural Art (当代社会与文化艺术),*" translated by Li Zhongze, published by Sichuan People's Publishing House, 2000 edition, page 41.
15. Chen Qilu: "*Study on Taiwan's Indigenous Culture (台湾土著文化研究),*" published by Taiwan Lianjing Publishing Co., Ltd. in 1992, page 7.
16. Darcy Ula Bay ・ Bima: "Taiwan's Indigenous Culture Series: Taiwan's Minority Bunun Tribe," published by Taiwan Indigenous Publishing House, 2003 edition, page 81.
17. Chen Yulan: "Taiwan's Minorities," published by Taiwan Hiking Culture Co., Ltd., 2004 edition, page 49.
18. Hong Yingsheng: "*Footprints of Taiwan's Indigenous People: Ten Tribes' Cultural Legends (台湾先住民脚印:十族文化传奇),*" published by Taiwan Times Culture Publishing Co., Ltd., 1993 edition, page 250.
19. Zhang Weizhen: "*A Study on the Deconstruction and Reconstruction of Taiwan's Minority Ethnic Textiles (台湾少数民族织物解构与重构之研究)*" (Doctoral Thesis), Institute of Design, Yunlin University of Science and Technology, Taiwan, 2011 Edition.

20. Chen Zhengzhe: "*Study on the Diamond Pattern of the Atayal Tribe (泰雅族菱形纹饰之研究)*" (Master's Thesis), Taichung Institute of Technology, Taiwan, Department of Commercial Design, 2008 edition.
21. Hong Zujin: "*Study on the Application of Symmetry in Minority Ethnic Costumes (对称在少数民族服饰之运用研究)*" (Master's Thesis), Taiwan University of Arts, Institute of Plastic Arts, 2007 edition.
22. Fang Limei: "*Study on the Art of Atayal Weaving (泰雅织布艺术之研究)*" (Master's Thesis), Fo Guang University of Humanities and Social Sciences, Taiwan, Institute of Art Studies, 2005 edition.
23. Zhou Jinhong: "*Study on the Traditional Craft Changes of the Atayal Tribe Beishi (泰雅族北势群传统工艺变迁之研究)*" (Doctoral Thesis), Taiwan Normal University, Institute of Industrial Education, 2001 edition.
24. Zhou Peijun: "*Study on the Inheritance Education of Atayal Weaving Craft: Taking Taoyuan County Fuxing Township as an Example (泰雅族织布工艺的传承教育之研究:以桃园县复兴乡区域为例)*" (Master's Thesis), Taiwan University of Arts, Department of Fine Arts, 2012 edition.
25. Wu Mengrong: "*I Weave, Therefore I Am: Cultural Production of Female Craft Weavers of the Truku Tribe in Xiulin Township, Hualien County (我织故我在:花莲县秀林乡太鲁阁族女性工艺编织者之文化生产)*" (Master's Thesis), Donghua University, Taiwan, Institute of Ethnic Relations and Culture, 2012 edition.
26. Zhang Weizhen: "*Study on the Deconstruction and Reconstruction of Taiwan's Minority Ethnic Textiles (台湾少数民族织物解构与重构之研究)*" (Doctoral Thesis), Yunlin University of Science and Technology, Taiwan, Design Institute, 2011 edition.
27. Su Qiufeng: "*Study on the Clothing Culture of the Paiwan Tribe: Taking the Lavavul Group's Sandimen Tribe as an Example (排湾族服饰文化之研究:以拉瓦尔群三地门部落为例)*" (Master's Thesis), Tajen University of Science and Technology, Taiwan, Institute of Leisure Health Management, 2010 edition.
28. Chen Zhengzhe: "*Study on the Diamond Pattern of the Atayal Tribe*" (Master's Thesis), Taichung Institute of Technology, Taiwan, Department of Commercial Design, 2008 edition.
29. Huang Weiwen: "*The Application of Traditional Totem Art of Taiwan's Minority Ethnic Groups in Wood Art Creative Design: A Case Study of the Puyuma Totem in Taitung (台湾少数民族传统图腾艺术应用于木艺创意设计:以台东卑南族图腾为例)*" (Master's Thesis), Tropical Agriculture and International Cooperation Research Institute, Pingtung University of Science and Technology, Taiwan, 2006 Edition.
30. Guo Yumin: "*The Commercialisation of Sandimen Paiwan Clothing and Its Changing Significance (三地门排湾衣饰商品化及其意义变迁)*" (Master's Thesis), Institute of Anthropology, Tsinghua University, Taiwan, 2006 Edition.
31. Huang Zhaomin: "*A Study of the Mythology and Legends of Tattooing Among Taiwan's Minority Ethnic Groups (台湾少数民族文身神话传说研究)*" (Master's Thesis), Department of Chinese Literature, Soochow University, Taiwan, 2005 Edition.
32. Luo Shuxing: "*A Study on the Changes and Formation of the Wedding Attire of the Rukai Tribe in Wutai: A Case Study of the Ba Family Wedding (雾台鲁凯族婚礼盛装的变迁与形制构成之研究:以巴家婚礼为个案分析)*" (Master's Thesis), Department of Textile and Clothing, Fu Jen Catholic University, Taiwan, 2004 Edition.
33. Xue Huiying: "*New Styles of Paiwan Women's Clothing: An Exploration of Their Form and Content (排湾族女性服饰之新款式:以其形式与内容作探究)*" (Master's Thesis), Institute of Applied Design, Shu-Te University, Taiwan, 2004 Edition.
34. Li Zhaoyu: "*A Study on the Origin of the Form of Shao Tribe Clothing (邵族服装之形制溯源研究)*" (Master's Thesis), Institute of Art, National Cheng Kung University, Taiwan, 2004 Edition.
35. Chen Shuqiao: "*A Study on the Relationship Between Bunun Tribe Textile Clothing and Identity: A Case Study of the Kadu Tribe in Zhuo Community (布农族织品服饰与认同关系之研究:以卓社群卡度部落为例)*" (Master's Thesis), Institute of Art, National Cheng Kung University, Taiwan, 2004 Edition.

36. Zhou Jinhong: "*A Study on the Changes in the Traditional Crafts of the Atayal Tribe in Beishi (泰雅族北势群传统工艺变迁之研究)*" (Doctoral Thesis), Institute of Industrial Education, National Taiwan Normal University, 2001 Edition.
37. Wang Xiaojun: "*A Study on the Weaving Tools of the Saisiyat Tribe of Taiwan's Minority Ethnic Groups (台湾少数民族赛夏族编器研究)*" (Master's Thesis), Institute of Art, National Cheng Kung University, Taiwan, 2001 Edition.
38. Zhou Peijun: "*A Study on the Inheritance Education of the Atayal Tribe's Weaving Craft: A Case Study of the Fuxing Township Area in Taoyuan Country (泰雅族织布工艺的传承教育之研究:以桃园县复兴乡区域为例)*" (Master's Thesis), Department of Fine Arts, National Taiwan University of Arts, 2012 Edition.
39. Zhu Zhongling: "*A Study on the Clothing Culture of the Pepohoan: Taking the Pepohoan Clothing Artifacts in Taiwan Museums as an Example (平埔族服饰文化研究:以台湾博物馆平埔族服饰文物为例)*" (Master's Thesis), Museum Studies and Antiquities Conservation Research Institute, Tainan University of the Arts, Taiwan, 2012 Edition.
40. Hong Zujin: "*A Study on the Use of Symmetry in Minority Ethnic Costumes (对称在少数民族服饰之运用研究)*" (Master's Thesis), Institute of Plastic Arts, Taiwan University of the Arts, 2007 Edition.
41. Huang Zhaomin: "*A Study on the Tattoo Myths and Legends of Taiwan's Minority Ethnic Groups (台湾少数民族文身神话传说研究)*" (Master's Thesis), Department of Chinese Literature, Soochow University, Taiwan, 2005 Edition.
42. Chen Shuqiao: "*A Study on the Relationship between Bunun Tribe Textile Clothing and Identity: Taking the Kadu Tribe of Zhuo Community as an Example (布农族织品服饰与认同关系之研究:以卓社群卡度部落为例)*" (Master's Thesis), Institute of Art Studies, National Cheng Kung University, Taiwan, 2004 Edition.
43. Zhu Zhongling: "*A Study on the Clothing Culture of the Pepohoan: Taking the Pepohoan Clothing Artifacts in Taiwan Museums as an Example*" (Master's Thesis), Museum Studies and Antiquities Conservation Research Institute, Tainan University of the Arts, Taiwan, 2012 Edition.
44. Zhang Weizhen: "*A Study on the Deconstruction and Reconstruction of Taiwan's Minority Ethnic Textiles*" (Doctoral Thesis), Institute of Design, Yunlin University of Science and Technology, Taiwan, 2011 Edition.
45. Chen Zhengzhe: "*A Study on the Diamond Patterns of the Atayal Tribe (泰雅族菱形纹饰之研究)*" (Master's Thesis), Department of Commercial Design, Taichung Institute of Technology, Taiwan, 2008 Edition.
46. Huang Zhaomin: "*A Study on the Tattoo Myths and Legends of Taiwan's Minority Ethnic Groups*" (Master's Thesis), Department of Chinese Literature, Soochow University, Taiwan, 2005 Edition.
47. Luo Shuxing: "*A Study on the Changes and Formulation of the Wedding Dress of the Rukai Tribe in Wutai: A Case Study of the Ba Family Wedding (雾台鲁凯族婚礼盛装的变迁与形制构成之研究:以巴家婚礼为个案分析)*" (Master's Thesis), Department of Textile and Clothing, Fu Jen Catholic University, Taiwan, 2004 Edition.
48. Wang Xinying: "*A Study on the Totem Stories of the Paiwan and Rukai Tribes (排湾族与鲁凯族图腾故事研究)*" (Master's Thesis), Institute of Children's Literature, Donghua University, Taiwan, 2004 Edition.
49. Guo-Bin Huang: "*An Attempt to Classify the Diamond Patterns in the Traditional Textiles of Taiwan's Indigenous Peoples (台湾原住民族伝統織物における菱形文様の分類を試みる)*" (An attempt to classify the diamon patterns in the traditional textiles of Taiwan's minority ethnic groups), https://reurl.cc/XVvzRE, 23rd October 2022.
50. He Yingzi: "*The Development of Color Recognition among the Indigenous Peoples of Southern Taiwan*" (The development trend of color recognition among the minority ethnic groups in southern Taiwan), in "History of Life Culture," Issue 59, 2011.
51. Haruna Tamoto: "*A Study on the Representation and Production of 'Indigenous Crafts': A Case Study of the Revival of Indigenous Textiles in Taiwan (「原住民工芸」の表象と制作をめぐる一考察:台湾原住民の織物復興を事例に)*" (A study on the representation and

production of "minority ethnic crafts": A case study of the revival of minority ethnic textiles in Taiwan), in "History," Issue 81, 2012.
52. Junko Higashimura: "*Ethnoarchaeology of the Circular Waist Loom: From the Weaving Techniques of Taiwan's Indigenous Peoples (輪状式腰機の民族考古学：台湾原住民の機織技術から)*" (Ethnoarchaeology of the circular waist loom: From the weaving techniques of Taiwan's minority ethnic groups), in "Japan Shunei Taiwan Minority Ethnic Research Association: Taiwan Minority Ethnic Research," Issue 12, 2008.
53. Riki Chijiwa: "*Sutaro Chijiwa and the Study of Indigenous Taiwanese Architecture: Research Anecdotes (台湾山地原住民住居に関する研究：ルカイ族及びパイワン族の伝統家屋における「祖柱」紋様について)*" (Sutaro Chijiwa and the study of Taiwanese minority ethnic architecture: Research anecdotes), in "Bulletin of the Takaoka University of Law," Issue 23, 2012.
54. Kiyoshi Tanaka et al.: "*A Study on the Dwellings of the Indigenous Peoples in the Mountainous Areas of Taiwan: On the 'Ancestor Pillar' Patterns in the Traditional Houses of the Rukai and Paiwan Tribes (台湾山地原住民住居に関する研究:ルカイ族及びパイワン族の伝統家屋における「祖柱」紋様について)*" (A study on the Dwellings of the minority ethnic groups in the mountainous areas of Taiwan: On the "ancestor pillar" patterns in the traditional houses of the Rukai and Paiwan tribes), in "Bulletin of Tokyo Kasei Gakuin University," Issue 39, 1999.
55. Guo-Bin Huang: "*An Attempt to Classify the Diamond Patterns in the Traditional Textiles of Taiwan's Indigenous Peoples (台湾原住民族伝統織物における菱形文様の分類を試みる)*" (An attempt to classify the diamond patterns in the traditional textiles of Taiwan's minority ethnic groups), https://reurl.cc/XVvzRE, 23rd October 2022.
56. Zeng He Yingzi: "*The Development of Color Recognition in the Indigenous People of Southern Taiwan*" (The Development Trend of Color Recognition in the Minority Ethnic Groups of Southern Taiwan), published in "History of Life and Culture," Issue 59, 2011.
57. Li Rengui: "*The Ethnic Groups and Migration of the Austronesian People in Taiwan (台湾南岛民族的族群与迁徙)*" (Revised Edition), Taiwan Vanguard Publishing, 2011, pages 76–77.
58. Lin Xiuche: "*The Correct Naming of the Seediq Tribe (赛德克族正名),*" commissioned by the Taiwan Minority Ethnic Groups Committee, Taiwan Law and Politics University Minority Ethnic Groups Research Centre, 2007, page 80.
59. Liao Shouchen: "*The Social Organisation of the Atayal Tribe (泰雅族的社会组织),*" Taiwan Tzu Chi University of Medicine and Humanities, 1998.
60. Taipei University Folklore and Anthropology Research Room: "*Research on the System of Taiwan's Minority Ethnic Groups: Volume One (台湾少数民族系统所属之研究:第一册本文篇),*" translated by Yang Nanjun, Taiwan Minority Ethnic Groups Committee, Taiwan Nantian Bookstore Co., Ltd., 2011, page 111.
61. Chen Qilu: "*Ethnicity and Culture (民族与文化),*" Taiwan Dawn Cultural Enterprise Co., Ltd., 1981, page 39.
62. Fei Luoli: "*Exploration of the Culture and Language Classification of Taiwan's Indigenous Tribes (台湾土著族的文化、语言分类探究),*" Taiwan Central Research Institute Ethnology Research Institute, 1994, page 438.
63. Daxi Wulawan Bima: "*Taiwan Indigenous Culture Series: Taiwan's Minority Ethnic Group, the Atayal Tribe (台原文化丛刊系列:台湾的少数民族泰雅族),*" Taiwan Indigenous Publishing, 2001, page 208.
64. Suzuki Sakutaro: "*Research on Taiwan's Indigenous Tribes (台湾の蕃族研究),*" Taiwan Nantian Bookstore Co., Ltd., 1988, page 37.
65. Lin Daosheng: "*Minority Ethnic Group Myths: Complete Collection of Stories (少数民族神话:故事全集)*" (Volume Two), Taiwan Han Art Color Research Cultural Enterprise Co.,Ltd., 2001, page 29.
66. Mori Ushinosuke: "*Atlas of the Living Conditions of the Indigenous People in Taiwan during the Japanese Occupation (日据时期本省山地同胞生活状况图集),*" translated by Huang Yaodong, Taiwan Province Literature Committee, 1983, page 4.

67. Zosan Yōkichi, Ōnishi Yoshitomo: "*Collection of Raw Barbarian Legends (生蕃伝说集)*," Nantian Bookstore Co., Ltd., Taiwan, 1996 edition, pages 74–75.
68. Liao Shouchen: "*The Social Organisation of the Atayal Tribe (泰雅族的社会组织)*," Tzu Chi University of Medical and Humanities, Taiwan, 1998 edition.
69. Daxi Wulawan · Bima: "*Atayal Myths and Legends (泰雅族神话与传说)*," Morning Star Publishing Co., Ltd., Taiwan, 2002 edition, page 40.
70. Taiwan Old Custom Survey Association: "*Barbarian Investigation Report: Volume 5, Atayal Prequel (蕃族调查报告书：第五册泰雅族前篇)*," translated by the Institute of Ethnology, Academia Sinica, Taiwan, 2012 edition, pages 289–290, 298.
71. Li Yiyuan: "*Images of Culture: Observations on Religion and Ethnic Culture (文化的图像:宗教与族群的文化观察)*" (Volume II), Yun Chen Cultural Industries Co., Ltd., Taiwan, 1992 edition, page 194.
72. Daxi Wulawan · Bima: "*Taiwan Original Culture Series: Taiwan's Minority Atayal Tribe (台原文化丛刊系列:台湾的少数民族泰雅族)*," Taiwan Original Publishing, 2001 edition, page 233.
73. Hei Dai · Bayan: "*Exploring the Lifestyle of the Atayal People: An Atayal's Personal Testimony (泰雅人的生活型态探源:一个泰雅人的现身说法)*," Hsinchu County Cultural Affairs Bureau, Taiwan, 2002 edition, page 36.
74. Liu Qiwei: "*Art Anthropology: Primitive Thinking and Creation (艺术人类学:原始思维与创作)*," Lion Books Co., Ltd., Taiwan, 2002 edition, page 159.
75. Lin Huixiang: "*Cultural Anthropology (文化人类学)*," Commercial Press, Taiwan, 2007 edition, page 313.
76. Ni Jianlin: "*The Origin of Decoration: A Study of Primitive Decorative Art (装饰之源:原始装饰艺术研究)*," Chongqing University Press, 2007, p.69.
77. Taiwan Old Custom Survey Association: "*Survey Report on the Customs of the Aboriginal Tribes: Volume One, Atayal Tribe (番族惯习调查报告书:第一卷泰雅族)*," translated by the Institute of Ethnology, Academia Sinica, Taiwan, 1996 edition, p.81.
78. Josiane Cauquelin, "*The Aborigines of Taiwan: The Puyuma: from headhunting to the modern world*," London & New York: Routledge Curzon, 2004, P.218.
79. 76. Chen Chi-lu, "*Material Culture of the Formosan Aborigines*," Taipei: Southern Materials Center, Inc., 1998, p.100.
80. Cai Yushan: "*Textiles of Taiwan's Minority Ethnic Groups: Analysis of Weave Structure and Patterns (台湾少数民族织物:织纹结构与图案分析)*," Taichung County Cultural Center, Taiwan, 2006 edition, p.12.
81. Okamura Yoshiemon, Zhang Yongxin: "*Taiwan's Aboriginal Cloth (台湾の蕃布)*," Japan Yushodo Co., Ltd., 1969 edition, p.30.
82. Suzuki Takeshi: "*Customs of Taiwan's Minority Ethnic Groups (台湾少数民族风俗)*," translated by Wang Meijing, Taiwan Aboriginal Culture Publishing House, 1999 edition, p.79.
83. Li Shali: "*The Clothing Culture of the Paiwan Tribe (排湾族的衣饰文化)*," Taiwan Independence Evening News Cultural Publishing Department, 1993 edition,p.5.
84. Hu Jiayu, Cui Yilan: "*Study on the Collection of the Department of Anthropology, National Taiwan University (台大人类学系伊能藏品研究)*," National Taiwan University Press, 1998 edition, pp.112–130.
85. Taiwan Old Custom Survey Association: "*Survey Report on the Customs of the Aboriginal Tribes: Volume One, Atayal Tribe (番族惯习调查报告书:第一卷泰雅族)*," translated by the Institute of Ethnology, Academia Sinica, Taiwan, 1996 edition, pp.81–82.
86. Youlan Duoyou: "*Atayal Weaving Shadows (泰雅织影)*," Taiwan Daoxiang Publishing House, 2004 edition, pp.22–23.
87. Glen Dudbridge, "*Aborigines of South Taiwan in the 1880s*," Taipei: Shung Ye Museum of Minorities, Institute of Taiwan History, Academia Sinica's Taiwan, 1999, P.166.
88. Taiwan Old Custom Survey Association: "*Investigation Report of the Barbarian Tribes: Volume 7, The Atayal Tribe, Part Two (蕃族调查报告书：第七册泰雅族后篇)*," translated by the Institute of Ethnology, Academia Sinica, Taiwan, 2010 edition, page 150.

89. Taiwan Old Custom Survey Association: "*Investigation Report of the Barbarian Tribes: Volume 4, The Seediq and Truku Tribes (蕃族调查报告书：第四册赛德克族与太鲁阁族)*," translated by the Institute of Ethnology, Academia Sinica, Taiwan, 2011 edition, pages 94, 92.
90. Taiwan Old Custom Survey Association: "*Investigation Report of the Barbarian Tribes: Volume 5, The Atayal Tribe, Part One (蕃族调查报告书：第五册泰雅族前篇)*," translated by the Institute of Ethnology, Academia Sinica, Taiwan, 2012 edition, pages 270–271.
91. Li Yiyuan, Shi Lei, Ruan Changrui, Yang Fufa: "*The Atayal People of Klesan: Ethnographic Field Research and Study, Volume One (南澳的泰雅人:民族学田野调查与研究上册)*," Institute of Ethnology, Academia Sinica, Taiwan, 1963 edition, page 8.
92. Li Rengui: "*The Tribes and Migrations of Taiwan's Austronesian Peoples (台湾南岛民族的族群与迁徙)*," Taiwan Folk Culture Enterprise Co., Ltd., 1997 edition, page 178.
93. Li Jiyu: "*The Patterns and Colors of the Atayal Tribe's Traditional Clothing in Taiwan (台湾少数民族泰雅族方衣服饰的纹饰与色彩)*," in "Colorful Clouds: A Special Exhibition of Taiwan's Minority Clothing Culture," Taiwan History Museum, 2000 edition, pages 53–54.
94. Jiang Shaoying: "*Taiwan Pavilion at the Sino-Japanese Weaving Craft Exchange Exhibition (中日编织工艺交流展・台湾馆)*," Taichung County Cultural Centre, Taiwan, 1998 edition.
95. Li Shali: "*Taiwan's Minority Clothing Culture: Tradition, Meaning, Illustration (台湾少数民族衣饰文化:传统・意义・图说)*," Taiwan South Sky Book Co., Ltd., 1998 edition.
96. Shi Cuifeng: "*Body Decorations and Clothing of Taiwan's Minorities (台湾少数民族身体装饰与服饰)*," Taiwan History Museum, 2004 edition.
97. Mitsui Hideki: "*The Structure of Beauty (美の构成学)*," Japan Chuo Koron Shinsha, 1996 edition, page 104.
98. Zuo Hanzhong: "*Chinese Folk Art Modelling (中国民间美术造型)*," Hunan Art, 1994 edition, pages 293–294.
99. Wu Shan: "*Chinese Arts and Crafts Dictionary (中国工艺美术辞典)*," Taiwan Lion Book Co., Ltd., 1991 edition, page 108.
100. Zhang Tianxing, Wu Xiaobing, Guo Luyan: "*Decorative Patterns: Non- representational Graphic Art (装饰图案:非再现性的图形艺术)*," Nanjing Normal University Press, 2009 edition, pages 2–4.
101. Ye Liu Tianzeng: "*Chinese Ornamentation Research (中国纹饰研究)*," Taiwan South Sky Bookstore Co., Ltd., 1997 edition, page 13.
102. Peter F. Stone, "*Tribal and Village Rugs: The Definitive Guide to Design, Patterns, and Motifs,*" London: Thames & Hudson, 2004, P.332.
103. Huang Guobin, Zeng Heyingzi, Cai Yushan, etc.: "*Comparative Report on the Structure of Asian Design Seen in the Weaving Techniques and Pattern Colors of Traditional Textiles of Taiwan's 13 Indigenous Tribes (台湾原住民13族の伝統的織物の織制作技術と文様色彩にみるアジアンデザインの構造比較報告書)*" (Research on the Structural Comparison of Traditional Textile Weaving Techniques and Pattern Colors of Taiwan's Minority Tribes), Huang Guobin Research Room, Kobe University of Art and Design, Japan, 2012 edition, page 125 (unpublished report).
104. Daxi Wulawan Bima: "*Taiwan Indigenous Culture Series: Taiwan's Minority Atayal Tribe (台原文化丛刊系列:台湾的少数民族泰雅族)*," Taiwan Indigenous Publisher, 2001 edition, pages 225–226.
105. Taiwan Old Customs Survey Association: "*Survey Report on the Customs of the Tribes: Volume One, Atayal Tribe (番族惯习调查报告书:第一卷泰雅族)*," translated by the Institute of Ethnology, Academia Sinica, Taiwan, published by the Institute of Ethnology, Academia Sinica, Taiwan, 1996, p.255.
106. Peter F. Stone, "*Tribal and Village Rugs: The Definitive Guide to Design, Patterns and Motifs,*" London: Thames & Hudson, 2004, p.24.
107. Zhuge Kai: "*Ten Lectures on Design Art (设计艺术学十讲)*," Shandong Pictorial Publishing House, 2006 edition, p.158.

108. Lin Chonghong: "*Basic Design (基础设计)*," Taiwan Greater China Book Company Limited, 1997 edition, p.113.
109. Yang Qingtian: "*Theoretical Basis of Formative Design: Introduction to Formative Principles (造形设计的理论基础:造形原理导论)*," Taiwan Art Wind Hall Publishing House, 1996 edition, p.100.
110. Wang Wuji, Liang Juting: "*Basics of Graphic Design (平面设计基础)*," Taiwan Artists Publishing House, 1995 edition, p.46.
111. Song Ying, Liu Baoyue: "*Graphic Composition Design (平面构成设计)*," China Architecture & Building Press, 2005 edition, p.51.
112. Wang Wuji: "*Principles of Graphic Design (平面设计原理)*," Taiwan Lion Book Co., Ltd., 1996 edition, pp.16–18, 17.
113. Wachsler, Buschick Snyder: "*Design Basics (设计基础)*" (Volume II), translated by Lv Jingxiu, Taiwan Liuhe Publishing House, 1996 edition, p.344.
114. Teng Xuemei, Huo Kang: "*Pattern Design Teaching and Application (图案设计教学与应用)*," Higher Education Press, 2009 edition, p.45.
115. William Lidwell, Kritina Holden, Jill Butler: "*Universal Principles of Design (设计的法则)*," translated by Lv Hengying, Taiwan Origin Publishing House, 2008 edition, p.136.
116. Ernst Hans Gombrich, "The Sense of Order: A Study in the Psychology of Decorative Art," London: Phaidon Press Ltd., 1984, P.142.
117. Lin Shuyao: "*Basic Morphology (基本造形学)*," Taiwan Weixin Bookstore Co., Ltd., 1981 edition, pages 491–493, 494.
118. Tsuji Hiroshi, Sugiyama Akihiro: "*Morphology of Form (造形形態論)*," Japan Sanko Bookstore, 1985 edition, page 225.
119. Terry Weissman Knight, "*Transformations of Languages of Designs,*" (Thesis Ph. D. - University of California, USA: Ann Arbor, MI. UMI., 1986), P.78.

Chapter 3
Analysis of Traditional Textiles of the Pan-Taiya Ethnic Group

3.1 Research Approach

The systematic study of Taiwan's ethnic minorities began between 1895 and 1945. During that time, it went through stages such as the Tokyo Anthropological Society, the Temporary Taiwan Old Customs Survey Society, and the establishment of the Folk Ethnology Chair at Taipei Imperial University. The collection and investigation of related ethnographic specimens, such as Torii Ryūzō, Inō Kanori, and Mori Ushinosuke, have left many valuable materials for future generations. In terms of the research on the system to which Taiwan's ethnic minorities belong, such as Yiduanchizi Zang, Miyamoto Nobuto, and Ma Yuandong, they have laid the foundation for the classification and naming of ethnic minorities. The research foundation of Taiwan linguistics and the investigation and research of Austronesian languages, such as Ogawa Naoyoshi and Asai Keilun, are pioneers in this field. The research findings of the above anthropologists are of great importance in terms of material, ethnic group, and language. After 1945, with the establishment of the Institute of Ethnology of the Taiwan Central Research Institute, research on Taiwan's ethnic minority issues continued such as Chen Qilu, Li Yiyuan, Yang Nanjun, and other scholars who have also conducted research and discussion on different levels of the Paiwan and Atayal ethnic groups.

The study of ethnic minorities is inextricably linked with anthropological knowledge. Michael Howard mentions anthropology, a science that studies humans using scientific methods, trying to explain humans' various commonalities and differences by examining the biological and cultural aspects of the human past and comparing the differences between existing human societies [1]. Conrad Kottak also mentions the branches of anthropology, such as "cultural anthropology," for studying human society and culture through description, analysis, and interpretation and then explaining the similarities and differences between human society and culture. "Archaeological anthropology" studies various material remains, reconstructing, describing, and interpreting past human behavior and cultural patterns.

P. H. Wang and J. Li, *Interpreting the Evolution of Atayal Tribal Textile Patterns*,
https://doi.org/10.1007/978-981-96-5451-2_3

"Physical anthropology" is the study of human biological diversity in multiple times and spaces, based on the concern for biological variation. "Linguistic anthropology" studies language phenomena in the context of social culture, that is, across time and space [2]. Li Lixin also mentions in the book "Research Methods in Design Art" that the field investigation method of design uses anthropological methods to study design phenomena. In anthropological research, fieldwork is the most basic method [3].

Anthropologists, such as Inō Kanori, explore the field of specific human group culture through "ethnology",[1] mainly starting from the level of material culture, emphasizing the particular phenomena or representative objects seen in life customs, including the current state of life, industry, rituals, and language, etc. [4, 5]. Ethnology is based on comparative data, not only from ethnography but also from other sub-fields of anthropology. Wang Hailong also mentions other sub-fields of anthropology, such as visual anthropology, which is the study of images in anthropology, including anthropology's study of primitive art, artists, indigenous handicrafts, and those who make folk items themselves [6]. In addition to analyzing physical products, it is also a primary task to examine their cultural connotations through image research, using image and media culture analysis methods to study national spirit, a notable development in visual anthropology in recent years. Li Lixin also points out that art anthropology is a direct, complete, and comprehensive description of the characteristics and forms of specific art cultures. For the anthropological study of design, design culture should be regarded as the center of understanding human design activities, focusing on the research of design culture phenomena on the dissemination of design, comparison of design, and design values [7].

Anthropology not only engages in the recording and collecting of ethnographies and comparative studies of ethnology but also contributes fundamentally and irreplaceably to the deep understanding of human behavior, thought systems, and unique lifestyles. The specimens brought back from the field, whether placed in research institutions or concentrated in museums, are not merely playing the role of "cultural witnesses." Anthropology also attempts to answer ordinary, precious, simple, or complex objects and arts, further implying social culture's deep structure and inherent meaning [8]. Looking at the anthropological research on the material culture of Taiwan's minority ethnic groups, most of them only describe the appearance of objects without delving into the analysis of the structural textures on the surface of the objects, such as "Taiwan Barbarian Annals" [9] "Taiwan's Barbarian Cloth" [10], "Research on the Primitive Art of Taiwan's Indigenous Races"[11], "Research on Taiwan's Barbarian Tribes" [12] and other works. The content focuses on investigating the living conditions of minority ethnic groups, including social culture, material culture, and important events. Still, the part about clothing culture only describes the characteristics of costumes, needing a deeper analysis of the patterns

[1] Cultural anthropologists engage in two types of research: ethnography (based on fieldwork) and ethnology (based on cross-cultural comparison) to study and interpret cultural diversity. Cited from Conrad Kottak: "Cultural Anthropology: Exploring Cultural Diversity," translated by Xu Yucun, Taiwan Kui Crown Book Co., Ltd., 2005 edition, page 11.

of textiles. Suppose we want to clarify the complex artistic structure of material culture further. In that case, we must combine modern research techniques, such as digitally reconstructing the subtle features of patterns and analyzing the clues missed in previous research with the help of other research methods.

In the context mentioned above, this study uses the perspective of anthropology as the methodological basis. Still, due to the detailed analysis of textiles, it adopts the "content analysis" method to analyze the shape and color of textile patterns. It refers to the original records of cultural relics to understand the characteristics of the research samples' subgroups. Content analysis is an analysis of social products, exploring and understanding things or events that have existed in the past, also known as literature analysis or information analysis method [13]. Content analysis is a widely accepted method of text research. When conducting research, a set of categories is first established, then each category's frequency is calculated [14]. In summary, this study falls within the scope of qualitative research; therefore, it focuses on overall induction, mainly understanding the relationship between design matters and cultural essence and inferring its potential design rules.

Taking the Atayal tribe's textiles as the research object, the research samples are taken from museums and private collections at home and abroad. Although these textiles are all traditional costumes of the Atayal tribe, they come from different environments. The social significance implied by the cultural relics may vary due to differences in the historical background, leading to different cultural cognitions. At this time, if the historical axis is used for research analysis, it is difficult to determine the time point of the cultural relics accurately, and the time gap may lead to an inability to connect. Furthermore, the pattern of textiles is influenced by the tribe's distribution, which may lead to differences in the weaving patterns and even different forms. The color of the textiles may change with time and space and may be different from the appearance at the time. Currently, the color comparison method can record the color range. Therefore, the study of Atayal's traditional costumes can be analyzed from the perspective of design art, using the thinking that constitutes design. In addition to exploring tribal art and aesthetic forms, interpretations of patterns and colors can also be proposed.

3.2 Research Steps

This study is based on anthropological viewpoints combined with relevant knowledge of visual design, such as graphic composition and color theory. Through description, pattern, and cultural comparison, the "style standard" of Atayal textiles is identified in,[2] thereby understanding its material cultural characteristics [15]. The research process can be divided into the following four stages:

[2] Xie Shizhong believes that the methods for identifying cultural circles mainly include "the criterion of quality, or form," and "the criterion of quantity." The former refers to similar cultural items in two distant regions. To meet the conditions of formal relevance, it must be confirmed on the basis that the parties involved cannot logically produce the item. The quantity criterion emphasizes that various

1. Sample survey: When selecting Atayal textiles as research samples, consider the historical background of the cultural relics and whether the patterns and colors presented by the textiles are unique, such as ceremonial costumes and teaching costumes (for education and inheritance). Another is based on the original collection location to identify whether it can be classified as the same subgroup as the essential condition for selecting research samples. Suppose the cultural relics are severely damaged and cannot be identified, the repetition of the weave pattern is high, or it is difficult to determine the tribal system to which the object belongs. In that case, it is excluded from the research samples. When classifying samples, to ensure the correctness of the essential data registration, it is supplemented by "Clothing of Taiwan's Minorities: Tenri University Affiliated Tenri Reference Museum Collection" [16], "The Beauty of the Original: Mr. Chen Chengqing's Collection of Taiwan's Minority Artifacts"[17], "Research on the Collection of the Department of Anthropology, National Taiwan University" [18], "Taiwan's Minority Clothing Culture: Tradition · Meaning · Illustration"[19], "Shunyi Taiwan Minority Museum Collection Catalogue." [20]"Portrait and Commentary Collection of Taiwan's Minority Ethnic Groups"[21], "Taiwan Prehistoric Culture Museum Collection: Catalogue of Minority Ethnic Textiles and Ornaments"[22], "Reproducing Atayal: Catalogue of Traditional Atayal Clothing Reproductions"[23], "Investigation of the Takasago Tribe: Overview of the Barbarian Society"[24], "Research on the System of Taiwan's Minority Ethnic Groups: Volume One" [25] and other books serve as evidence, along with the digital collections of the Taiwan Central Research Institute of Ethnology[3] and the Atayal cultural relics digital collection database[4] to examine the background information of each research sample.

   The craft information and traces left by research specimens should be thoroughly inspected before analysis and recorded entirely and faithfully. The recording method can use text accompanied by images—the more detailed the recorded information, the more complete the information obtained for future research. During inspection and recording, first observe visually, using tools such as tape measures, magnifying glasses, color cards, cameras, etc. The inspection and recording of textiles can be divided into three categories: cultural relic information, material craft, and preservation status. Cultural relic information provides basic background information about the textile, including size, weave structure, etc.; material craft summarizes the materials and weaving techniques used in the textile production; preservation status examines whether the cultural

---

forms constitute a similar complex or organic whole, which can be used as evidence. Cited from Xie Shizhong: "Macro Exploration of Ethnic Anthropology: A Collection of Essays on Taiwan's Minorities," National Taiwan University Press, 2004 edition, pages 156–157.

[3] Taiwan Digital Archives, https://digitalarchives.tw, 2nd April 2023.

[4] Digital Archives of the Institute of Ethnology, Academia Sinica, Taiwan, https://ianthro.ioe.sinica.edu.tw, 2nd April 2023.

relic specimen can be listed as a research sample. Use a table to record the survey items, including object name, object type, collection location, appearance description, sample depiction, size, era, weave structure or pattern, material, primary color, etc. After completing the initial sample survey and archiving the research samples, the content of the information includes tribe name, object name, survey location, survey time, number, object explanation (such as material, technique, pattern, appearance, explanation, remarks), main pattern, primary color, detailed organization. This study obtained 73 samples for research use from domestic and foreign museums and private collectors, as shown in Fig. 3.1 and Appendix 1.

2. Pattern Analysis: Textile patterns all have a shape, and the shape characteristics of different textile patterns can be described in words, but it is challenging to present them accurately. Although photography can show a sense of reality, it cannot accurately reflect the specific structure of each part of the textile. Therefore, this study records the physical objects through text descriptions, photographic images, and pattern mapping to reproduce the appearance of textile patterns as realistically as possible. The weaving pattern is depicted using Pattern Maker for cross stitch software in conjunction with Adobe Illustrator drawing software. The former calculates the pattern based on the distribution of warp and weft grid lines, so when depicting the pattern, it must be constructed point-by-point to replace the traditional graph paper depiction method, as shown in Fig. 3.2. The latter is used because some textile patterns are not woven but sewn onto the fabric surface as decorations, so drawing software is used. Most textile patterns are colorful and complex, so only their unit patterns are selected for depiction to facilitate research and analysis. The notation for textile patterns in the text is sample number - front/back/pattern number/subgroup (textile type), such as NO.2- front ⓐ Makanji (men's sleeveless long robe). To analyze the structural elements of the pattern, it is necessary to fully understand the principles of textile weaving and the characteristics of textile patterns. The weaving craft of ethnic minorities is an additive concept, using the arrangement of warp and weft lines to expand and generate patterns. The organizational structure, i.e., the pattern layout method, is also the inherent form of the overall pattern. Here, the process of analyzing the pattern can use the concept of subtraction, removing the outer structure of the pattern layer by layer, thereby clarifying the core pattern. On the other hand, through the method of interconnecting graphics, such as separate configuration, repetitive configuration, intersection configuration, connection configuration, etc., the overall arrangement of the pattern can be analyzed, and the distribution of the woven structure can be observed.

Fig. 3.1 Types of Atayal textile clothing (compiled from this study)

Fig. 3.2 Pattern Maker for cross stitch software (drawn by this study)

3. Color Comparison: When conducting a color comparison, in addition to considering the universality and commonality of the textile industry, the reflective color conditions are also considered. Moreover, PANTONE is one of the global common indicators for color systems and color research. After the survey, the color data can be converted into RGB or CMYK values, which is beneficial for subsequent studies or other design applications. Therefore, the PANTONE FASHION + HOME cotton planner (2007 edition) compares textile colors. In addition, the Corel DRAW Graphics Suite X6/PANTONE FASHION + HOME cotton selector drawing software contains a color card database, which can correspond the data of the paper color card to RGB or CMYK values, which can be used when writing research results to indicate the color in the text, to record the color range of the textile. The operation steps are as follows:

a. Load the PANTONE FASHION + HOME cotton selector color card database.
b. Correspond the colors of the compared textiles to PANTONE color codes.
c. Convert to RGB or CMYK color values.
d. Indicate the colors of the textiles, with the primary color name as the range, such as black series (colorless), red series (colored), etc. This description method can also reduce the gap in color perception.

4. Statistical Analysis: The textiles were collected for research purposes, and apart from the total number of research samples from various subgroups of the Atayal tribe, most of the data was not in numerical form. Suppose we want to understand the correlation between patterns and color combinations through the color performance of textiles and the overall color combination performance of subgroup textiles. In that case, we must first encode the composite stripes formed by their patterns and color combinations. Next, we convert the determined data content into countable numbers to provide data for the following analysis step. Here, we use descriptive statistics to analyze the color performance of composite stripes in textiles. According to the book "Social Statistics and Data Analysis," descriptive statistics is the most incredible tool for providing a summary overview of the data, presenting the complete picture of sample characteristics concisely. When describing the overall profile of the sample, the focus is usually placed on four aspects of sample characteristics: frequency distribution, central tendency, dispersion tendency, and distribution shape [26]. Therefore, hue analysis and statistical analysis of color combinations for composite stripes can also be used to understand the structure and color relationships of the weave and provide explanations. The operation steps are as follows:

   a. Occurrence of composite stripe colors: Analyze the hues of composite stripes in textiles and count the number of single composite stripe color combinations. Taking the single piece of fabric in Table 3.1 as an example, the composite stripe colors appear most frequently in white (9), pink (8), orange (8), and yellow (7), followed by black (5), red (5), grey (2), blue (1). The maximum number of composite stripe color combinations is six: black, white, red, pink, orange, yellow; grey, white, pink, orange, yellow, blue; black, white, red, pink, orange, and yellow. Next, three colors: black, pink, yellow; black, white, orange; white, pink, orange. Five colors: grey, white, red, pink, yellow; white, red, pink, orange, yellow. Two colors: white and orange.
   b. Percentage of composite stripe colors: After counting the number of composite stripe color combinations, divide the total of each hue by the total number of occurrences to obtain the distribution ratio of textile pattern colors. For example, in Table 3.1, the total number of composite stripe colors in the entire piece of textile is 45 (total of both sides), and the number of

Table 3.1 Analysis of the colors of the compound stripes of the textile (using NO.39 women's long-sleeved dress as an example)

Item Name: Women's long-sleeved dress	Main colors: red, crange

Front of fabric garment/Back of fabric garment

Composite stripe color combination

Pattern number	Black	Gray	White	Red	Pink	Orange	Yellow	Blue	Sum
Front ⓐ	●		●	●	●	●	●		6
Front ⓑ			●			●			2
Front ⓒ	●				●		●		3
Front ⓓ	●		●			●			3
Front ⓔ			●		●	●			3
Front ⓕ		●	●		●	●	●	●	6
Back ⓐ		●	●	●	●		●		5
Back ⓑ	●		●	●	●	●	●		6
Back ⓒ			●	●	●	●	●		5
Back ⓓ	●		●	●	●	●	●		6
Occurrence	5	2	9	5	8	8	7	1	45
Ratio (%)	11.1	4.4	20	11.1	17.8	17.8	15.6	2.2	100

occurrences of white in the composite stripes is 9. Divide this data by the total number of colors; the calculation formula is 9 (white) / 45 (total colors) = 0.2 or 20% (percentage of occurrences). White accounts for 20%; pink accounts for 17.8%; orange accounts for 17.8%; yellow accounts for 15.6%; black accounts for 11.1%; red accounts for 11.1%; grey accounts for 4.4%; blue accounts for 2.2%.

c. Compound Stripe Color Combination Percentage: To understand the distribution of compound stripes' number and color combinations, we can group those with the same number of color combinations. Through statistics, we can obtain the proportion of color combinations and thus understand the characteristics of the color combinations presented by the textile pattern. Taking the single piece of textile in Table 3.2 as an example, the compound stripe color combinations include two-color, three-color, five-color, and six-color combinations. The calculation formula is four (the number of times white appears in the six-color combination) / 24 (subtotal of quantity) = 0.1666, which is 16.7% (percentage of occurrence). From this calculation, we can obtain the proportion of the color combination in the whole, such as the compound stripes of the six-color combination with white (16.7%), pink (16.7%), orange (16.7%), yellow (16.7%) as the colors with higher usage rates, followed by black (12.5%), red (12.5%), grey (4.2%), blue (4.2%). This analysis can also help understand the color performance of the textile pattern.

3.3 Research Process

This study focuses on the form and color of the Atayal tribe's textile patterns, understanding the design performance of the textile patterns through sample analysis. The research process is divided into thinking, concept, execution, analysis, and conclusion, as shown in Fig. 3.3. The research motivation, purpose, scope, and framework are established first. Then, through the collection, collation, induction, and interpretation of literature reviews, the content includes important literature and related theories and the understanding of the Atayal tribe's material culture, which is the basic knowledge for analyzing Atayal textiles. In addition, through the visual design's form language, the composition form of the textile patterns is understood, followed by the explanation of the research samples and the gradual formulation of the phased research methods, including sample survey, pattern analysis, color comparison, and statistical analysis. The sample survey establishes the essential data required for the research; pattern analysis and color comparison clarify the composition of the textiles; statistical analysis further explains the correlation between patterns and color matching. The exploration of textile design thinking mainly includes pattern form logic, compound stripe arrangement rules, and the association between patterns and colors, and through the origin and migration path of the sub-group legends, the geographical distribution of textile patterns is constructed. Finally, the research conclusions and subsequent suggestions are proposed based on the results of the analysis and interpretation.

Table 3.2 Analysis of color combinations of textile compound stripes (using NO.39 women's long-sleeved dress as an example)

Number of colors		Composite stripe color combination								Quantity Subtotal	Pattern Subtotal
		Black	Gray	White	Red	Pink	Orange	Yellow	Blue		
Two colors	Occurrence	0	0	1	0	0	1	0	0	2	1
	Ratio (%)	0	0	50	0	0	50	0	0	100	
Three colors	Occurrence	2	0	2	0	2	2	1	0	9	3
	Ratio (%)	22.2	0	22.2	0	22.2	22.2	11.1	0	100	
Five colors	Occurrence	0	1	2	2	2	1	2	0	10	2
	Ratio (%)	0	10	20	20	20	10	20	0	100	
Six colors	Occurrence	3	1	4	3	4	4	4	1	24	4
	Ratio (%)	12.5	4.2	16.7	12.5	16.7	16.7	16.7	4.2	100	
Sum	Occurrence	5	2	9	5	8	8	7	1	45	10
	Ratio (%)	11.1	4.4	20	11.1	17.8	17.8	15.6	2.2	100	

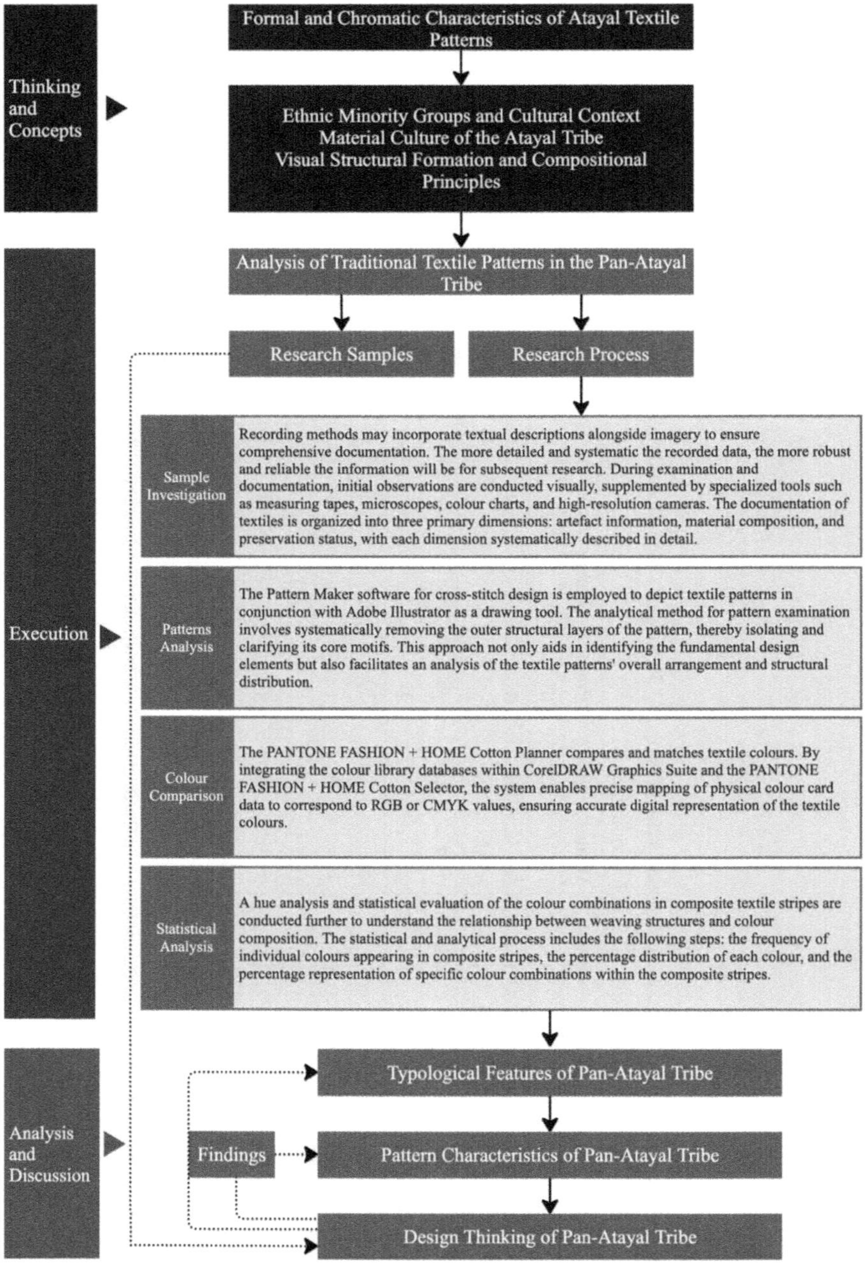

Fig. 3.3 Research process (drawn by this study)

References

1. Michael Howard: "*Cultural Anthropology (文化人类学),*" translated by Li Maoxing and Lan Meihua, Taiwan Hongzhi Cultural Enterprise Co., Ltd., 1997 edition, page 9.
2. Conrad Kottak: "*Cultural Anthropology: Exploring Cultural Diversity (文化人类学:文化多样化的探索),*" translated by Xu Yucun, Taiwan Kui Crown Book Co., Ltd., 2005 edition, pages 11–16.
3. Li Lixin: "*Research Methods in Design Art (设计艺术学研究方法),*" Jiangsu Fine Arts Publishing House, 2009 edition, page 240.
4. Conrad Kottak: "*Cultural Anthropology: Exploring Cultural Diversity (文化人类学:文化多样化的探索),*" translated by Xu Yucun, Taiwan Kui Crown Book Co., Ltd., 2005 edition, page 11.
5. Hu Jiayu, Cui Yilan: "*Research on the Ino Collection of the Department of Anthropology, National Taiwan University (台大人类学系伊能藏品研究),*" Taiwan University Press, 1998 edition, pages 40–41.
6. Wang Hailong: "*Visual Anthropology (视觉人类学)*" Shanghai Literature and Art Publishing House, 2007 edition, page 58.
7. Li Lixin: "*Research Methods in Design Art (设计艺术学研究方法),*" Jiangsu Fine Arts Publishing House, 2009 edition, pages 109–110.
8. Wang Songshan: "Museum Work and Ethnographic Field Methods *(博物馆工作与民族学田野方法),*" in Huang Zhenyan (ed.): "Collection and Museum of Folk/Ethnic Culture," Taipei University of the Arts Publishing Centre, 2011 edition, page 108.
9. Morimoto Chūnosuke: "*Taiwan Barbarian Records: Volume One (台湾蕃族志:第一卷),*" Nantian Bookstore Co., Ltd., Taiwan, 1917 edition.
10. Okamura Yoshiemon, Zhang Yongxin: "*Taiwan's Barbarian Cloth (台湾の蕃布),*" Yu Xiu Tang Co., Ltd., Japan, 1969 edition.
11. Sato Fumiichi: "*Research on the Primitive Art of Taiwan's Indigenous Races (台湾原住种族の原始艺术研究),*" Nantian Bookstore Co., Ltd., Taiwan, 1988 edition.
12. Suzuki Sakutarō: "*Research on Taiwan's Barbarian Tribes (台湾の蕃族研究),*" Nantian Bookstore Co., Ltd., Taiwan, 1988 edition.
13. Guan Xingsheng, Ruan Luyin, Wang Mingtang, et al.: "*Design Research Methods (设计研究方法),*" Quan Hua Books Co., Ltd., Taiwan, 2006 edition, page 60.
14. David Silverman: "*Interpreting Qualitative Data (解析质性研究法与资料),*" translated by Tian Zherong and Si Tu Yi, Weber Culture International Publishing Co., Ltd., Taiwan, 2010 edition, page 189.
15. Xie Shizhong: "*Macro Exploration of Ethnic Anthropology: A Collection of Essays on Taiwan's Minorities (族群人类学的宏观探索:台湾少数民族论集),*" National Taiwan University Press, 2004 edition, pages 156–157.
16. Tenri University: "*Costumes of Taiwan's Minorities: Collection of Tenri University's Tenri Reference Museum (台湾少数民族之衣饰:天理大学附属天理参考馆所藏),*" Tenri University, Japan, 1993 edition.
17. Editorial Committee of Taiwan History Museum: "*Beauty of Authenticity: Mr. Chen Chengqing's Collection of Taiwan's Minority Artifacts (原真之美:陈澄晴先生珍藏台湾少数民族艺术文物),*" Taipei City History Museum, Taiwan, 1997 edition.
18. Hu Jiayu, Cui Yilan: "*Research on the Ino Collection of the Department of Anthropology, National Taiwan University (台大人类学系伊能藏品研究),*" National Taiwan University Press, 1998 edition.
19. Li Shali: "*Taiwan's Minority Ethnic Clothing Culture: Tradition, Meaning, Illustration (台湾少数民族衣饰文化:传统・意义・图说),*" Nantian Bookstore Co., Ltd., Taiwan, 1998 edition.
20. Xu Yahui: "*Shunyi Taiwan Minority Museum Collection Catalogue (顺益台湾少数民族博物馆文物图录),*" Shunyi Taiwan Minority Museum, 1999 edition.
21. Dai Jialing: "*Taiwan's Minority Ethnic Portrait and Commentary Collection (台湾少数民族写真与解说集),*" Taiwan Avant-garde Publishing House, 2000 edition.

22. Fang Junwei: "*Taiwan Prehistoric Culture Museum Collection: Minority Ethnic Textiles and Jewellery Catalogue (台湾史前文化博物馆藏品:少数民族织品及饰品图录),*" Taiwan Prehistoric Culture Museum, 2002 edition.
23. Fang Junwei: "*Reappearance of Atayal: Pan-Atayal Traditional Costume Reproduction Catalogue (重现泰雅:泛泰雅传统服饰重制图录),*" Taiwan Prehistoric Culture Museum, 2008 edition.
24. Police Affairs Bureau, Aborigine Affairs Section: "*Investigation of the Takasago Tribe: Tribal Overview (高砂族调查书:蕃社概况),*" translated by the Institute of Ethnology, Academia Sinica, Taiwan, Institute of Ethnology, Academia Sinica, Taiwan, 2011 edition.
25. Folklore and Anthropology Research Room, Taipei University: "*Research on the System of Taiwan's Minority Ethnic Groups: Volume One Main Text (台湾少数民族系统所属之研究:第一册本文篇),*" translated by Yang Nanjun, Taiwan Minority Ethnic Committee, Nantian Bookstore Co., Ltd., Taiwan, 2011 edition.
26. Xie Xuzhou: "*Social Statistics and Data Analysis (社会统计与资料分析),*" Wiseman Cultural Enterprise Co., Ltd., Taiwan, 2008 edition, page 134.

Chapter 4
Characteristics of Pan-Atayal Group Textile Types

4.1 Tayal

Anthropologists divide the Atayal tribe into Atayal and Seediq, the former further distinguished into Seqoleq and Truku, the latter divided into east and west groups based on their residence on the east or west side of the Central Mountain Range in Taiwan [1]. The Atayal tribe has different names, such as Atayal, Tayal, Taiyal, etc. These tribal names are only used in the northwest of Taiwan and are self-proclaimed. However, in the southeastern Musha and Truku areas, they call themselves Sədeq, Sēdeq, or Səjeq. Regardless of the name, they all mean "people" [2].

According to anthropological classification, the research samples are divided into Atayal and Seediq, such as the Seqoleq of the Tayal, including the Fuxue, Makanji, Msbtunux, Klesan, Malepa, and Taranan; the Truku includes the Klesan, Plnawan, Bunsui, Beishi, Nanshi, and Meicyubus. The Seediq consists of the Musha, Pulibao, and Truku. This chapter focuses on the subgroups of the Tayal, describing their cultural background, geographical distribution, and textile characteristics, and presents the patterns and colors of Atayal textiles through mapping and comparison.

4.1.1 Textiles of the Hakul (Hakku)

The Hakul (Hakku) calls itself Hakul, named after its residence in the deep mountains. The Hakul (Hakku) live south of Baigu Mountain, along the Beigang River's upper reaches. It is composed of four communities: Mastbwan (Mastodon community), Tebilun (Tabilun community), Mknazi (Makanaji community), and Zihing. It borders the Mli'pa tribe (Malepa) to the east, the Bb'ala tribe (Mabarnor) to the west, reaches Baigu Mountain to the north and borders the territories of the Troku and Tawca' tribes (Seediq group) to the south [3, 4]. According to the description in the "Taiwan Minority Ethnic History Language Culture Dictionary," the early activity area of the

P. H. Wang and J. Li, *Interpreting the Evolution of Atayal Tribal Textile Patterns*,
https://doi.org/10.1007/978-981-96-5451-2_4

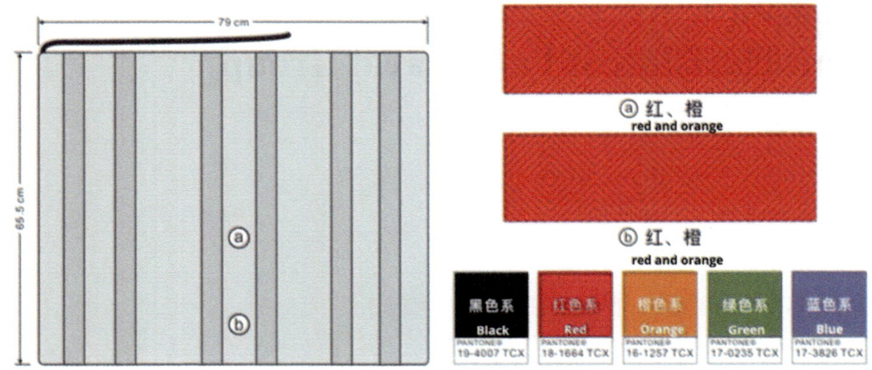

Fig. 4.1 Unisex Shawl (NO.01) (Drawn by this study)

Hakul (Hakku) of the Atayal people was within the Renai Township Fashang Village, starting from Kagawa River in the south and reaching Kimun River in the north. The Mastbwan (Mastodon, now called Ruiyan) tribe on the Sibayan plateau above the village originates in the Seqoleq legend. The Hakul (Hakku) has a long history and is the root of the Atayal tribe.[1]

Taking the NO.01 unisex shawl as an example, it is made from three pieces of fabric. The red part has a diamond pattern, with a tie on the upper left side. The whole piece has two types of composite stripes. Materials: ramie, cotton, wool. Techniques: twill weave, double warp pick-up. Patterns: diamond pattern, stripes. Colors: black, red, orange, green, blue. The composite stripes are composed of diamond patterns arranged horizontally. The patterns are not repeated in pairs but are similar in size, with different internal structures, as shown in Fig. 4.1ⓐ ⓑ. In Fig. 4.1ⓐ , an X-shaped structure is inserted between the two diamond patterns, and in Fig. 4.1ⓑ , the interior of the diamond pattern is composed of smaller diamonds that gradually form a larger diamond structure—a color combination of the patterns red and orange.

4.1.2 Makanji Textiles

The Makanji, self-proclaimed Makanji (Makanaji), lives in the northern part of Mount Daba, including the mountain tribes between the two Knaji streams, composed of eight communities. They are distant neighbors of the Mnibu, facing the Sykaru tribe to the west, bordering the Mrqwang (Mariluo) north, and looking south to Mount Daba [5]. According to the description in the "Encyclopedia of Taiwan's Minority History, Language and Culture," the Makanji lives in the uppermost stream of the Msbtunux River under the jurisdiction of the Gaogan branch of Taoyuan Prefecture,

1 Encyclopedia of Taiwan's Minority Ethnic History, Language, and Culture, https://reurl.cc/6Nky6O, 6th April 2023.

behind the Mrqwang (Mariluo). The nearest group, Tabaho, is about 14 miles southeast of Zhudong and 15 miles southeast of Msbtunux. The language system of the Makanji is like that of the Kao Bay group, with a slight influence on the Northern group. The tribes within the Knazi group include Tabaho (Tianpu), Chinsubo (Zhenxibao), Taiyakan (Taigang), Takejin, Haga, Saihowan (Saihouwan), Makalugiyahu (Jinlu), Sagayachin (Baishi) and so on. The ancestors of the Kinaji group, around 1600, moved from the Makanaji of the Hakul (Hakku) in Nenggao County, Taichung Prefecture, led by Butakaraho, with about 30 people, to the Chinshibo tribe under the county, located in the middle of the Knazi Mountain, and established the first community. After that, the descendants further divided and established new tribes.[2]

Taking the NO.02 men's sleeveless long robe as an example, the body of the robe is made from two pieces of hemp fabric. The main woven patterns are concentrated on the garment's front, middle, and lower back, with a large diamond pattern on the back. The whole piece has three types of composite stripes. Materials: ramie, wool. Techniques: plain weave (base), colored weft floating weave, colored weft pick-up sandwich weave. Patterns: diamond pattern, zigzag pattern, mountain pattern. Colors: black, grey, peach red, purple. The composite stripes are composed of small diamond patterns arranged in pairs in a continuous stripe structure, as shown in Fig. 4.2ⓐ . The large pattern on the back of the garment, as shown in Fig. 4.3ⓐ , is composed of alternating zigzag and mountain patterns. In Fig. 4.3ⓑ , the large diamond pattern is structured with an X-shape, consisting of four diamond patterns, with different diamonds on the top, bottom, left, and right—color combination of the patterns: grey, peach red.

4.1.3 Msbtunux Tribe Textiles

According to the Jiaobanshan of the Msbtunux, initially, Besu' Nagwi and Watan Temu, two tribal leaders, led their tribesmen from Papak Waqa (Mount Daba) down to the Msbtunux River (Dahan River) basin, establishing tribes in various places. Eventually, Besu' established today's Kkijay (Jiaobanshan, located in Zeren Village, Fuxing District, Taoyuan County), and Watan established today's Qara (Zhutoujiao, situated in Changxing Village, Fuxing District, Taoyuan County) [6]. The Msbtunux consists of ten communities: Kzyay (Kachia), Silung (Shilang), Hbun (Haben), Ulay (Wulai, now in Nanshi Village, Heping District, Taichung County), Gihing (Yixing), Rahu (Lahao), Qehuy (Kuihui), ZhuTouJiao, Sinazi (Sinaji), and Qoyow (Gaorao). "Msbtunux" is the name the Qing government gave, and the tribesmen call themselves Sbtunux [7].

Taking the NO.03 men's and women's shawl as an example, it comprises three types of fabric surfaces. The main fabric surface's stripes are woven in plain and zigzag patterns, and the three sides are decorated with double warp pick-up weaving.

[2] Encyclopedia of Taiwan's Minority Ethnic History, Language, and Culture, https://reurl.cc/6Nky6O, 6th April 2023.

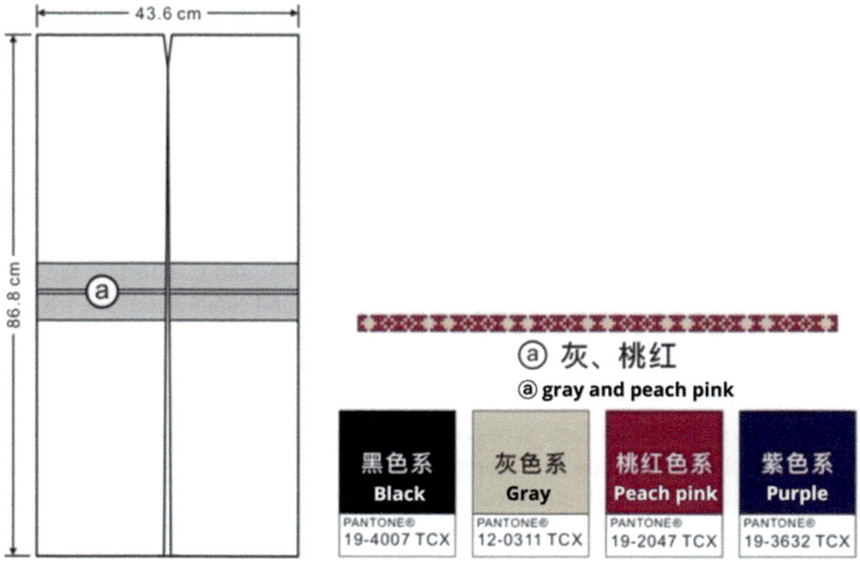

Fig. 4.2 Men's Sleeveless Long Robe—Front (NO.02) (Drawn by this study)

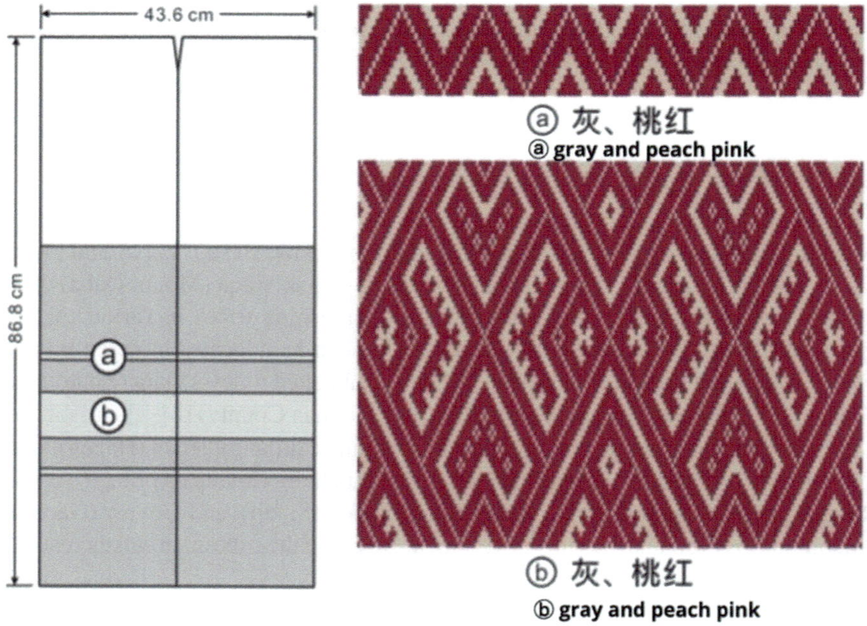

Fig. 4.3 Men's Sleeveless Long Robe—Back (NO.02) (Drawn by this study)

The whole piece has four types of compound stripes. Material: ramie, cotton, wool, plastic buttons. Techniques: warp-faced plain weave, small twill weave, special warp pick-up, embroidery (binding). Patterns: diamond pattern, zigzag pattern, mountain pattern, stripe. Colors: black, grey, red, pink, yellow, blue, brown. The compound stripes are diamond patterns interspersed with zigzag patterns, as shown in Fig. 4.4ⓐ . Figure 4.4ⓑ stripe organization is mainly a mountain pattern, with a twill structure interspersed in the middle. Larger patterns are shown in Fig. 4.4ⓒ ⓓ ; their compound stripes are composed of diamond patterns; the two patterns are almost similar but with slight differences; the × shape in the middle of the two diamond patterns is different. Color combinations of the patterns: black, yellow; black, grey; black, grey, brown. The rest of the Msbtunux textiles and pattern position numbers are shown in Fig. 4.5.

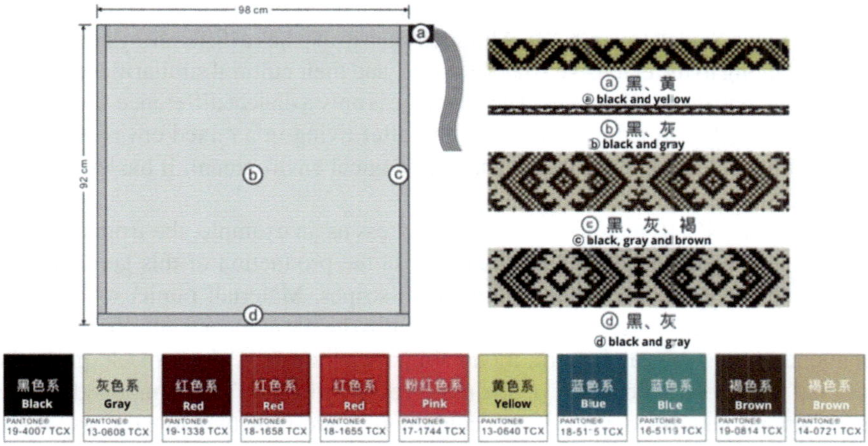

Fig. 4.4 Men's and Women's Shawl (NO.03) (Drawn by this study)

Fig. 4.5 Schematic of the Pattern Positions of the Msbtunux Textiles (Drawn by this study)

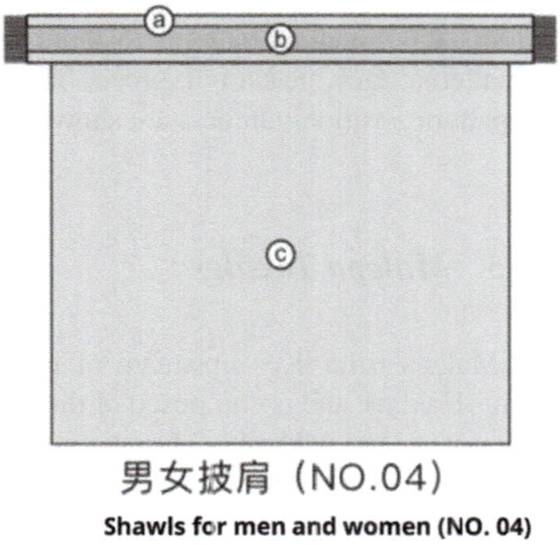

4.1.4 Klesan Textiles

The Klesan is named after the administrative region, roughly referring to the scope of the aboriginal land of Su'ao County, Taipei State (Klesan Township, Yilan County). There is no ethnic group in the system, including many outsiders. In the past, the tribes of the Klesan were almost all located on the slopes of the upper and middle reaches of the Ləliyung-Tolo (Da Zhuo River). The west, north, and south sides of their residential area are surrounded by mountain ridges, becoming a state isolated from other ethnic groups. The so-called Klesan itself includes four subgroups that call themselves Məbəala (west and east, such as Wuta Society, etc.), Kəna-Xaqul (middle, such as Jinyang Society, etc.), Tausa (south, such as Kumoyao Society, etc.), and Mənebo (central and slightly north, such as Hagabaris Society, etc.), each with its own fixed territory, customs, and dialects. In terms of the system, Kəna-Xaqul belongs to the Saikolik system; Məbəala and Mənebo belong to the Tseole; Tausa belongs to the Seideke system [8]. Although each group has a different system origin, they all belong to the extensive Atayal system, and their cultural similarities are more significant than their differences. The language is only a dialect difference. This group has undoubtedly formed a standard culture after living in a mixed environment for a long time and adapting to the standard ecological environment. It has established the concept of our group in psychology [9].

Taking the NO.10 men's sleeveless long dress as an example, the front and back are full of woven patterns. There are errors in the production of this garment, and the whole piece has ten kinds of compound stripes. Material: ramie, wool. Techniques: plain weave (base), sandwich weave, warp heavy plain pick flowers. Patterns: diamond pattern, zigzag pattern, mountain pattern, stripe. Colors: black, grey (the original color of ramie), red, peach red, green, and blue. The compound stripes are diamond patterns arranged in a continuous pattern of two squares, as shown in Figs. 4.6ⓐ and 4.7ⓐ ⓑ ⓒ ⓓ ⓔ ⓕ ⓘ . Figures 4.6ⓐ and 4.7ⓐ ⓑ ⓒ have similar diamond pattern structures, 4.7ⓓ ⓔ ⓕ ⓖ ⓗ ⓘ are complex patterns, among which 4.7ⓓ ⓔ are composed of smaller diamonds to form large patterns, 4.7ⓕ ⓘ are composed of larger single diamonds, and 4.7ⓖ ⓗ are consisting of zigzag patterns and mountain patterns to create compound stripes—color combinations of the patterns: grey, peach red, green, blue; black, grey. The rest of Klesan's textiles and pattern position numbers are shown in Figs. 4.8.1 and 4.8.2.

4.1.5 Malepa Textiles

The Malepa tribe lives upstream of Beigang Creek, in the northeast area of the Hakul (Hakku), and is composed of three communities: Muyaw (Moyao), Pelmwan (Bilumoan), and Mkbabu (Mogubabo). In the early days of the Muyaw Society's establishment, they, unfortunately, encountered a famine, and the tribe members had to eat bean pods (Kripa), which led other societies to mockingly call them Kripa,

Fig. 4.6 Men's Sleeveless Long Dress—Front (NO.10) (Drawn by this study)

which later was misinterpreted as Malepa (meaning high place) [10]. The Hakul (Hakku) and the Malepa are adjacent, with the Malepa initially separating from the Hakul (Hakku) to form its group. Both have their own Qotox-gaga organizations (originally meaning to maintain the same habits), so they are independent groups, but there are no significant differences in language and customs [11].

Taking the NO.26 tobacco bag as an example, the body of the bag is sewn into a slightly square-shaped long tube with a piece of flower-woven cloth, and straps are sewn on both sides of the upper end. The whole piece has four types of composite stripes. Material: ramie, wool. Technique: plain weave (base), colored weft floating weave, colored weft picked flower sandwich weave. Pattern: diamond pattern, mountain pattern. Colors: black, grey, white, red, yellow, green, blue, purple. The composite stripes are arranged in pairs continuously, as shown in Fig. 4.9ⓐ ⓑ ⓒ ⓓ . Figures 4.9ⓑ ⓓ are arranged in horizontal intervals, with different patterns inserted between the diamond patterns, namely diamond patterns and vertical stripes. Color combinations of the patterns: black, grey; black, grey, red; white, red; black, grey. The rest of the Malepa textiles and pattern location numbers are shown in Fig. 4.10.

4.1.6 *Taranan Textiles*

The Taranan tribe is scattered in the mountainous area upstream of Xindian Creek, composed of nine communities: Kara'muci', Tranan, Cyaqung, Tunok, Rahaw, Syoryaw (also known as Qnahan), Raga', Tampya' (also known as Kayu'), and Ulay (also known as Urai, now Wulai Village, Wulai Township, Taipei County). Among them, Kara'muci', Tranan, Cyaqung, Tunok, and Rahaw are collectively called Rmugan; Syoryaw, Raga', Tampya', and Ulay are collectively called Ulay.

Fig. 4.7 Men's Sleeveless Long Dress—Back (NO.10) (Drawn by this study)

The tribes collectively refer to these nine communities as Yawi Puna quux lung, meaning "the descendants of Yawi Puna living along the river [12]." Initially, the leader Yagwi Puna thought that the land of Pinsbkan (Binsibo Gan) was narrow and inconvenient for farming, so he led his tribesmen to cross Quri Sqabu (Si Yuan Ya Kou) ("Sqabu" saddle part), along the Da Keng Creek and then over the mountains to the present location. Yagwi had children including Yukan, Iban, Batu', Pajas, Tana, Pehu', Kaqui, Sapu', etc. Yukan went elsewhere, the descendants of Batu', Pehu', and Kaqui were cut off, and the remaining ones were Iban (the great-grandfather of the leader Batu' Watan), Sapu', Pajas, and the descendants of Tana. It is said that the descendants of these four people own the land of the whole tribe [13].

Taking the NO.29 men's and women's shawl as an example, it is made of three pieces of cloth, with tie loops on both sides of the upper end, and the whole piece has

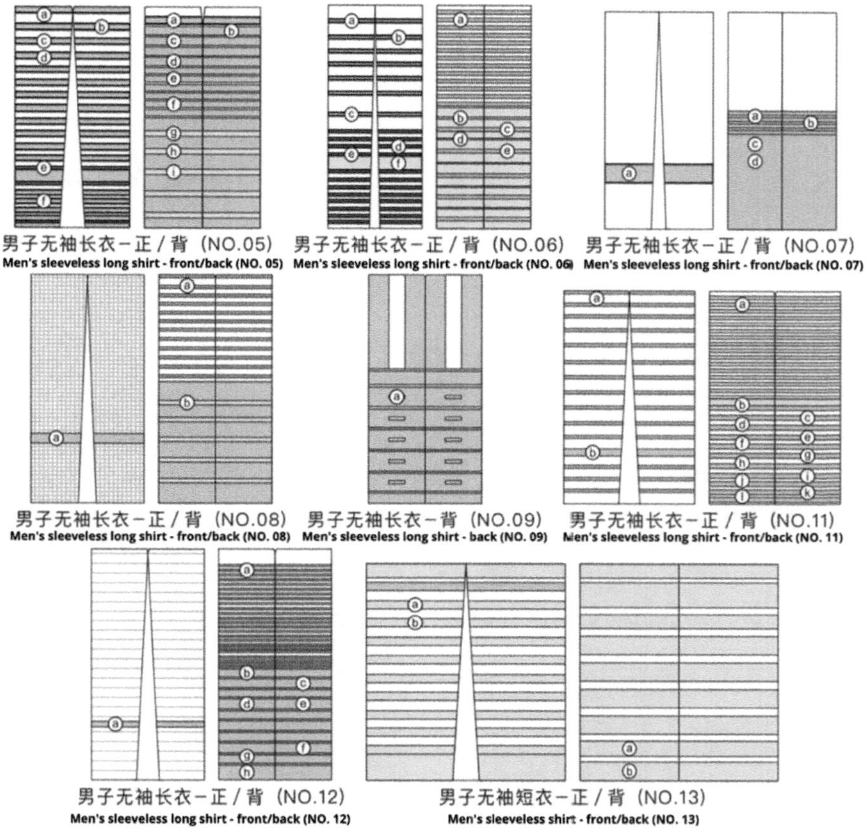

Fig. 4.8.1 Schematic of the Pattern Positions of the Klesan's Textiles (Drawn by this study)

only one type of composite stripe. Material: ramie, cotton, wool. Technique: plain weave (base), unique warp jacquard. Pattern: circular pattern, stripe (dot pattern). Color: black, white, peach red, blue. The composite stripe is composed of nearly circular patterns arranged in a continuous pattern of two sides, in which the circular patterns are arranged horizontally in an alternating manner, as shown in Fig. 4.11ⓐ—the color combination of the pattern: black and white. The rest of the ruler group woven products and pattern position numbers are shown in Fig. 4.12.

4.1.7 Plnawan Woven Products

The Plnawan tribe lives in the mountainous area along the Plnawan Creek, a tributary of the Zhuoshui River, composed of two communities, Cgi'as (Jiu Ge Si) and Sime'ul (Si Mi Ao Lu), a small group among the Atayal people [14]. The Plnawan is

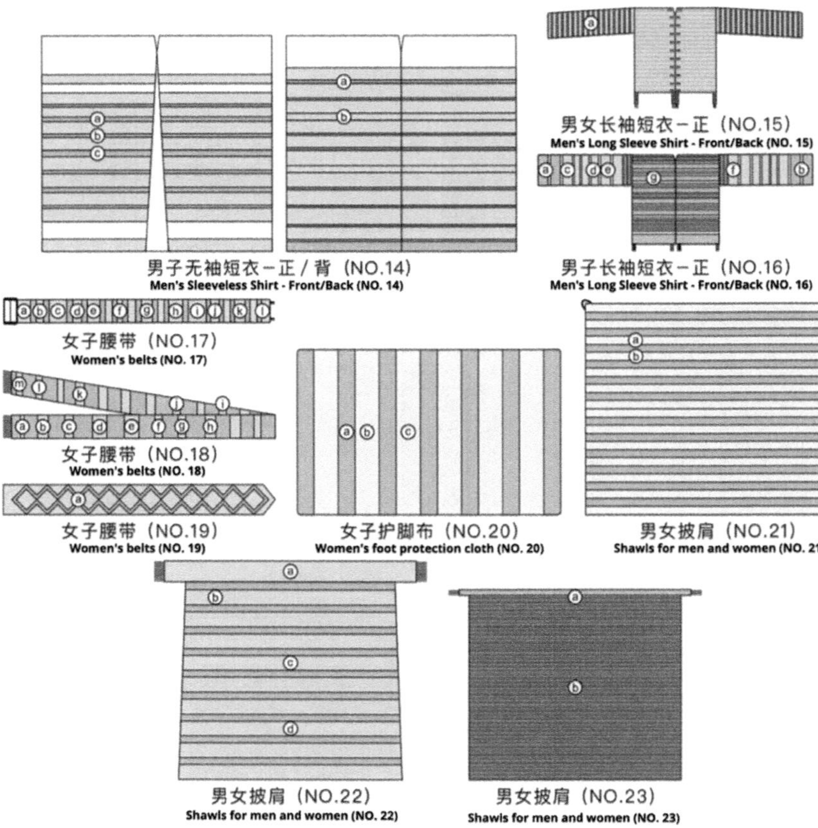

Fig. 4.8.2 Schematic of the pattern location of Klesan textiles (drawn by this study)

represented by the Plnawan (Perugawan), bordered by the Bunun tribe on the southwest side and surrounded by the Sədeq tribe on the northeast side. The Plnawan belongs to the Tseole of Atayale (self-proclaimed Itayal), so its geographical location is isolated. The residents of the Plnawan are composed of three small groups, one of which is the Perugawan, which was initially the residents of the Plnawan; the second is the Murauts, which originally lived in the Puli Basin and moved into the Plnawan 103, 104 years ago; the third is due to intermarriage, belonging to the Sədeq system's Dao Ze group, Tuo Lu Ge group, and a group that moved into the Plnawan from the far away Hakul (Hakku) [15].

Taking the NO.30 men's long-sleeved robe as an example, the body of the robe is made by joining two pieces of linen and then sewing the sleeves on both sides. The robe pattern is based on red, with blue and white sawtooth fine horizontal stripes interwoven. A large diamond pattern is woven below the waist on the back, and the collar is edged in dark blue. The whole piece has five types of composite stripes. Material: ramie, cotton. Techniques: plain weave (base), colored weft float weave,

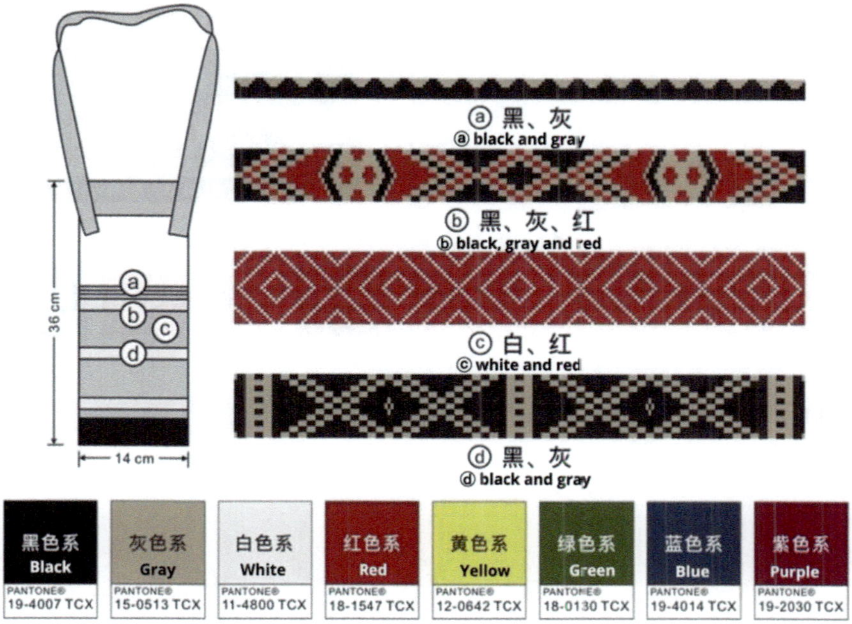

Fig. 4.9 Tobacco bag (NO.26) (drawn by this study)

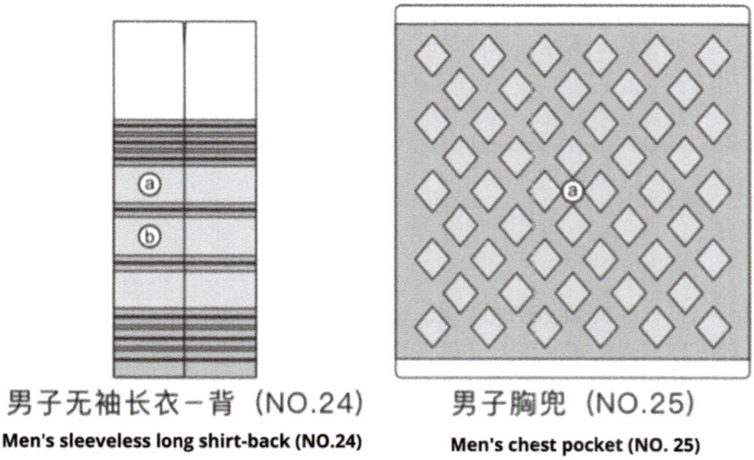

Fig. 4.10 Schematic of the pattern location of Malepa textiles (drawn by this study)

colored weft pick-up interweave. Patterns: diamond pattern, zigzag pattern, mountain pattern, cross pattern, stripe pattern. Colors: black, white, red, blue. The composite stripes are arranged in a continuous square pattern, as shown in Figs. 4.13ⓐ ⓐ and 4.14ⓐ ⓑ ⓒ . The structure of Fig. 4.13ⓐ is like Figs. 4.13ⓑ and 4.14ⓐ

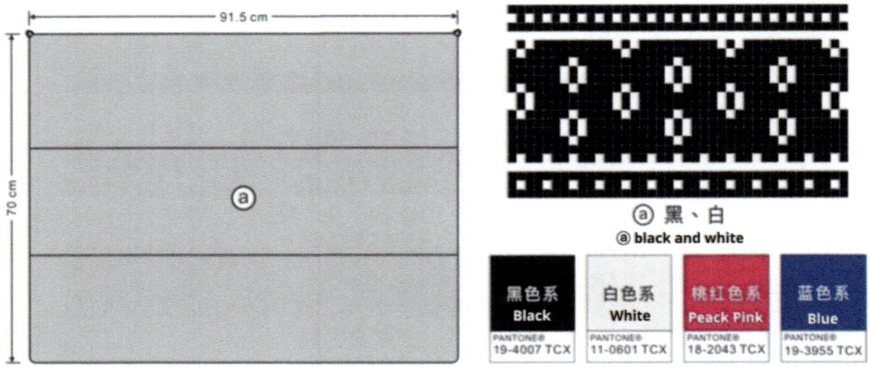

Fig. 4.11 Men's and Women's Shawl (NO.29) (Drawn by this study)

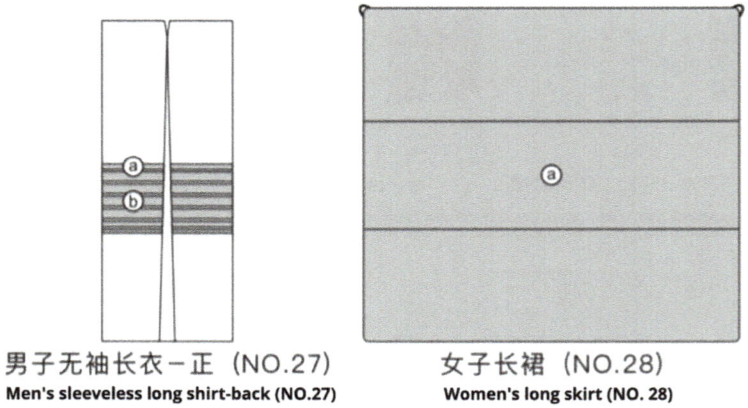

Fig. 4.12 Schematic of the pattern position of the ruler group woven products (Drawn by this study)

. Figure 4.14ⓑ is arranged with the upside-down mountain pattern, forming the zigzag pattern in the reversed white area. Figure 4.14ⓒ is a large diamond pattern arranged in a continuous square stripe pattern. Still, the space structure between the two diamond patterns is constructed by taking half of the large diamond pattern. Color combinations of the patterns: white, red, blue; white, blue; black, white; red, blue.

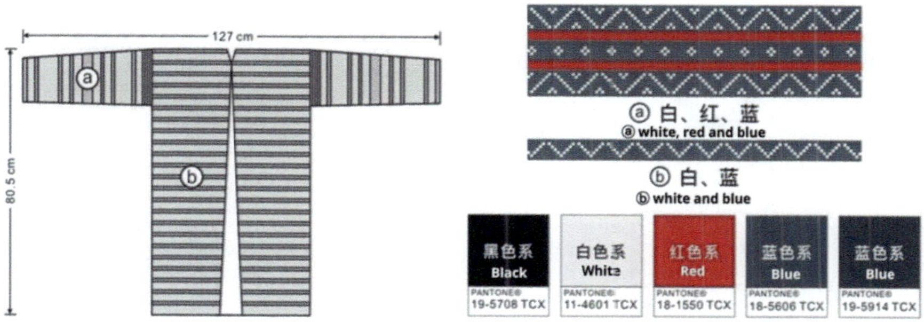

Fig. 4.13 Men's long-sleeved robe—front (NO.30) (Drawn by this study)

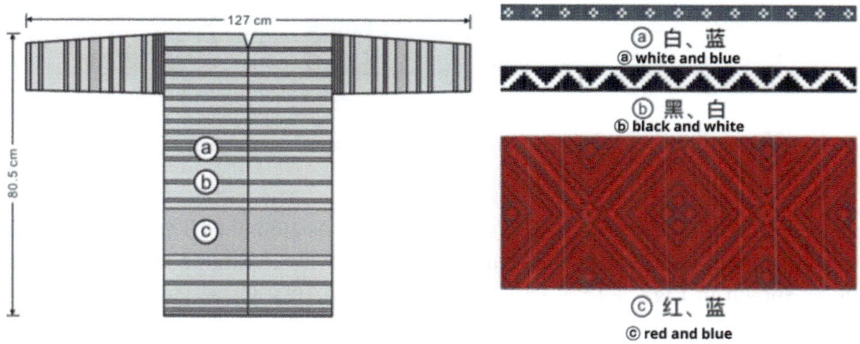

Fig. 4.14 Men's long-sleeved robe—back (NO.30) (Drawn by this study)

4.1.8 Bunsui Textiles

The Bunsui lives along the Bunsui and Xishui tributaries of the Da'an River, consisting of six communities: Pkwali' (Baguali), Tabiras (Dabulas), Ma'aw, Sahyan, Saburuk, and Sqawn. The territory of the Bunsui extends from the north of Meicyubus Mountain to the south of Jiali Mountain, to the Bajiaolin of the Dahu Sub-Hall, south to Yangmu Mountain, and northeast back to Meicyubus Mountain. It borders the Beishi tribe to the east, the Beishi and Dahu tribes to the south, the Minzhuang to the west, and the Seediq group to the north [16, 17].

Taking the NO.39 women's long-sleeved dress as an example, this style is a bridal dress. The upper half is sewn from two pieces of fabric, the lower half is made from different patterned fabric pieces, and the sleeves are joined from another two pieces of fabric. Black cotton fabric is sewn at the cuff position. The garment is entirely of woven patterns, with ten composite stripes. Materials: ramie, wool, cotton, pearl, metal. Techniques: plain weave (base), double warp pick-up, sandwich weave, embroidery. Patterns: diamond pattern, × pattern, stripe. Colors: black, grey, white, red, pink, orange, yellow, green, blue, and brown. The composite stripes are

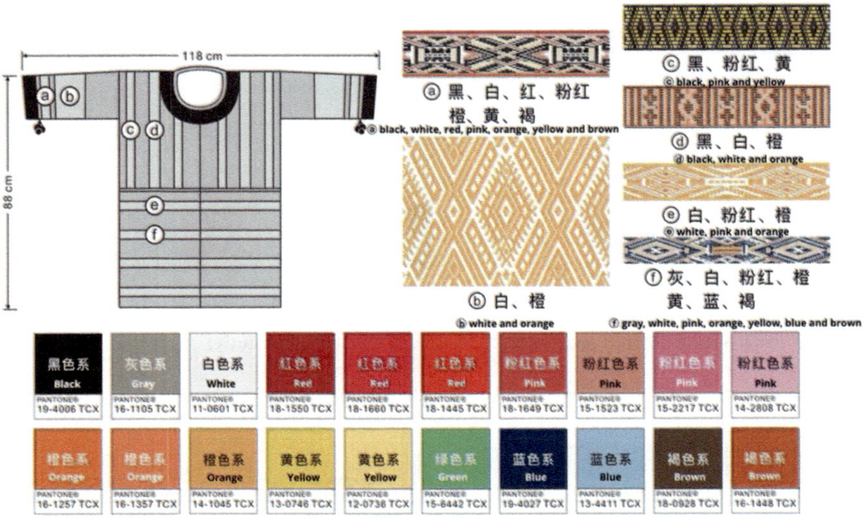

Fig. 4.15 Women's long-sleeved dress—front (NO.39) (Drawn by this study)

composed of diamond patterns, using dots and lines as elements. The pattern shapes are complex and highly variable, arranged in a two-way continuous stripe organization, as shown in Figs. 4.15ⓐ ⓑ ⓒ ⓓ ⓔ ⓕ and 4.16ⓐ ⓑ ⓒ ⓓ . Figure 4.16ⓒ separates its patterns with reverse white lines, dividing them into groups of different shapes, such as diamond patterns and × patterns. As shown in Fig. 4.15ⓑ , large diamond patterns are arranged in a left–right alternating pattern, and Fig. 4.16ⓐ is placed up and down, interspersed with a small diamond pattern. These two patterns are both composed of small diamond groups forming large organizations. Color combinations of the patterns: black, white, red, pink, orange, yellow, and brown; white, orange; black, pink, yellow; black, white, and orange; white, pink, and orange; grey, white, pink, orange, yellow, blue, and brown; grey, white, red, pink, and yellow; black, white, red, pink, orange, and yellow; white, red, pink, orange, and yellow. The remaining Bunsui textiles and pattern position numbers are shown in Figs. 4.17.1 and 4.17.2.

4.1.9 *Beishi Textiles*

The Beishi lives along the Da'an River and is known as Mə-Pənox. Like the Dahuhu, Bunsui, and Jialapai groups in the north, the Beishi also regard the Daba Tip Mountain as the ancestral birthplace. According to legend, the tribe members moved down the mountain along the Da'an River or Dajia River and finally settled in their current residence [18]. During the period from 1895 to 1945, the Beishi was composed of eight societies, namely Lobugo (Old House E Society), Suru (Sulu Society),

Fig. 4.16 Women's long-sleeved dress—back (NO.39) (Drawn by this study)

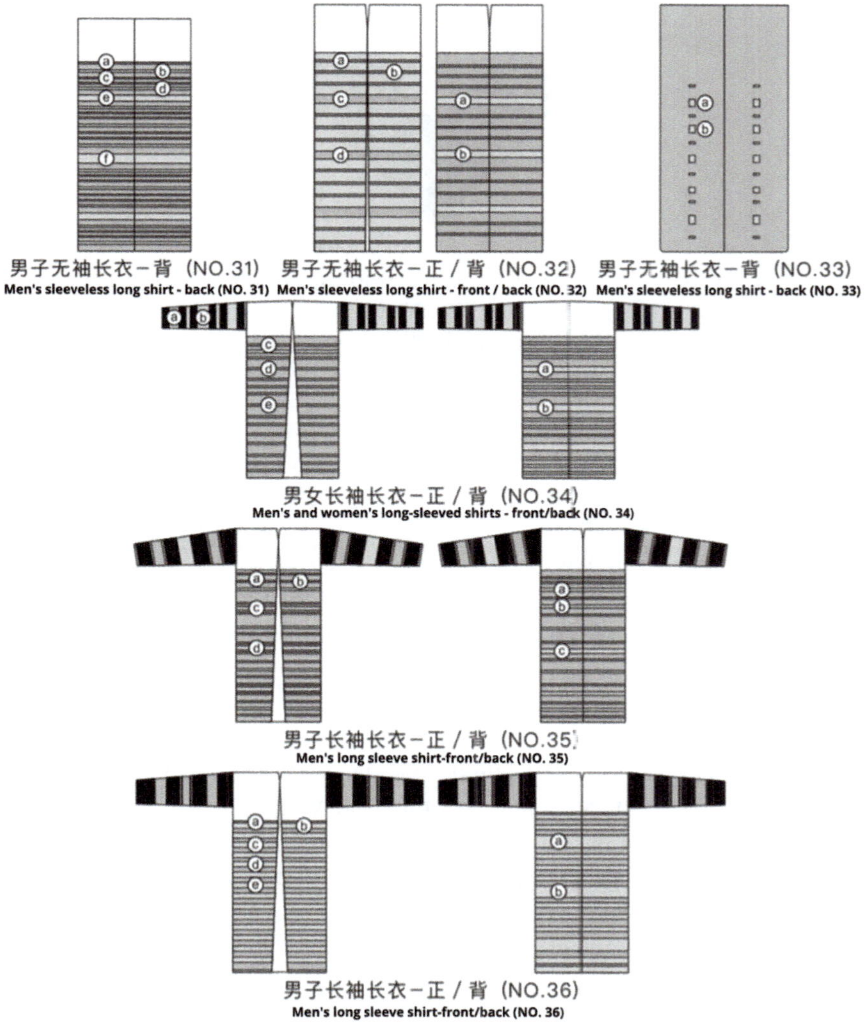

Fig. 4.17.1 Schematic of the pattern positions of the Bunsui textiles (Drawn by this study)

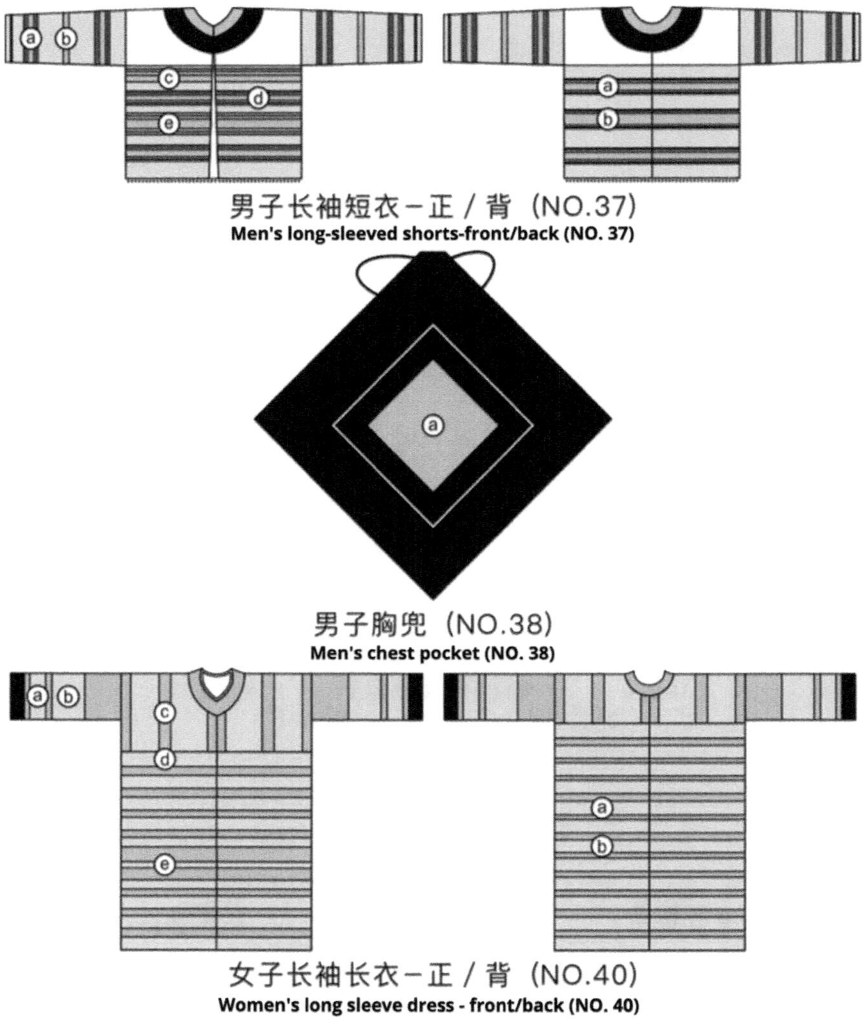

Fig. 4.17.2 Schematic of the pattern positions of the Bunsui textiles (Drawn by this study)

Mabrahaw (Mabihao), Cinmuy (End Tail), Temukbonay (Get Wood Mo Nai), Lobung (Lu Peng), Mnapang (Ma Na Bang), and Buyung (Wu Rong) [19]. The territory of the Beishi extends from the Daxueshan, northeast to the Slbija Mountain, north to the Daba Peaks, west to the Meicyubus Mountain, further southwest through the Cinmui Mountain, Sima Limit Mountain, and Manabang Mountain to the Guanyin Mountain, and from the southeast Yanzui Mountain (Dajian Mountain) back to the Daxueshan. In the far east, there are the Slamao (Shalamao) and Sqojao; to the south are the Nan Shi tribes; to the west are the villages under the Dahuhu Sub-Hall; and to the north are the Bunsui and Dahuhu [20].

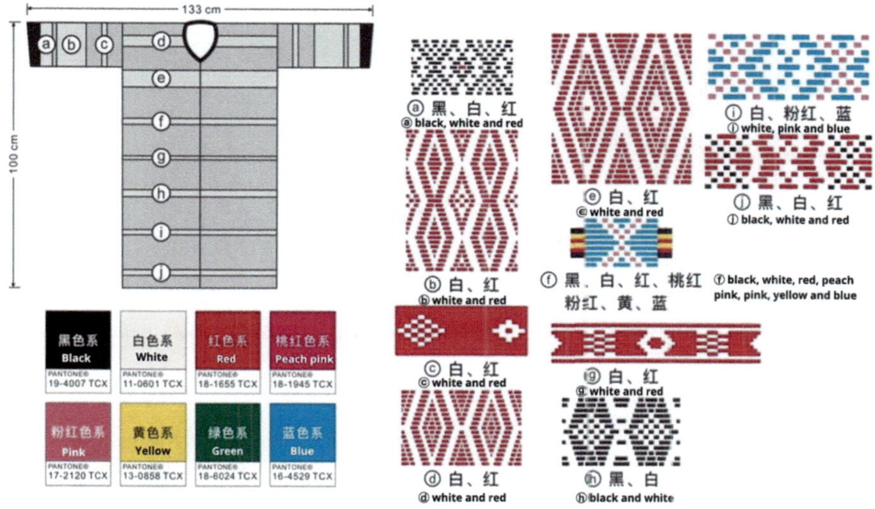

Fig. 4.18 Women's long-sleeved dress—front (NO.44) (Drawn by this study)

Taking the NO.44 women's long-sleeved dress as an example, this style is a bridal gown made by joining two pieces of woven fabric to form the body of the dress, and the sleeves are made by joining another two pieces of cloth, with black cotton fabric sewn at the cuff position. The garment is covered with woven patterns and 17 types of composite stripes. Material: Ramie, wool. Techniques: Plain weave (base), pick weave, sandwich weave. Patterns: Diamond pattern, × pattern, stripe. Colors: Black, white, red, peach red, pink, yellow, green, blue. The composite stripes are composed of diamond patterns, mainly points and lines, and are arranged in a square pattern, as shown in Figs. 4.18ⓐ ⓓ ⓔ ⓗ ⓘ ⓙ and 4.19ⓐ ⓑ ⓒ ⓓ ⓔ ⓖ. Large diamond patterns, such as Figs. 4.18ⓑ and 4.19ⓕ, are arranged repetitively, with another type of diamond pattern interwoven in the middle. In addition, the patterns in Figs. 4.18ⓒ ⓕ ⓖ are composed of diamond patterns, × patterns, and stripes to form composite stripes. Color combinations of the patterns: Black, white, red; white, red; black, white, red, peach red, pink, yellow, blue; black, white; white, pink, blue; black, white, red, pink, green; black, white, blue. The rest of Beishi's woven products and pattern position numbers are shown in Fig. 4.20.

4.1.10 Nan Shi Woven Products

The Nan Shi is just a geographical name. It is a collection of various elements, which can be divided into those belonging to the Saikolik (Bo'ai, Timburan, Qingshan, Ulai-ruma, Nan Aleng's Ulai-ləngawan and Ulai-Kaoyao), as well as the Shao Lai and Bai Mao, which seem to belong to the Beishi, and possibly the Aleng, which belongs to this small group [21]. The Nanshi is scattered around the middle reaches

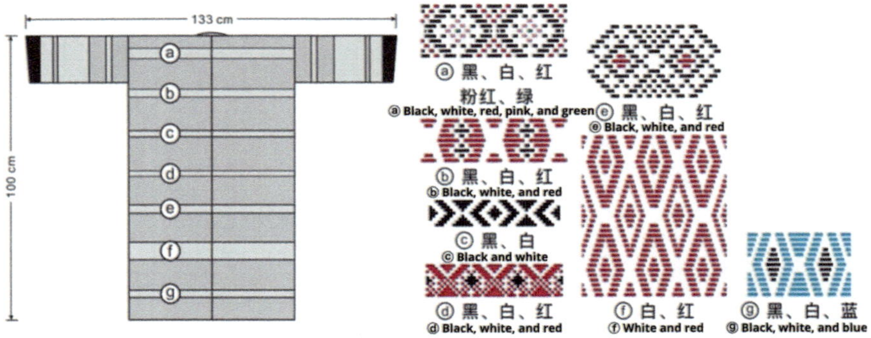

Fig. 4.19 Women's long-sleeved dress—back (NO.44) (Drawn by this study)

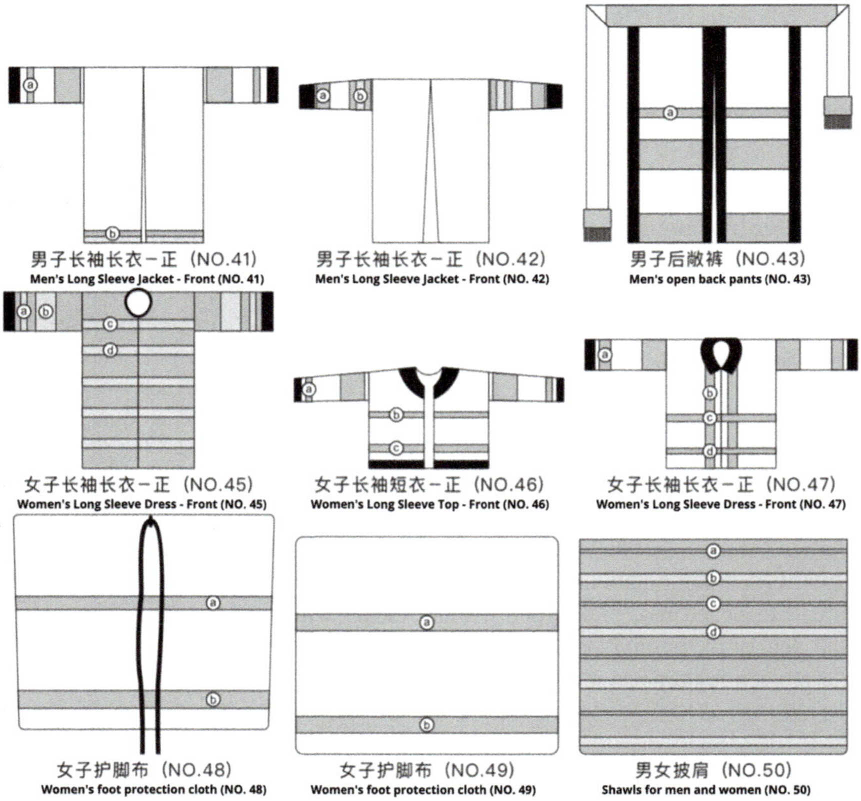

Fig. 4.20 Schematic of the pattern positions of Beishi's woven products (Drawn by this study)

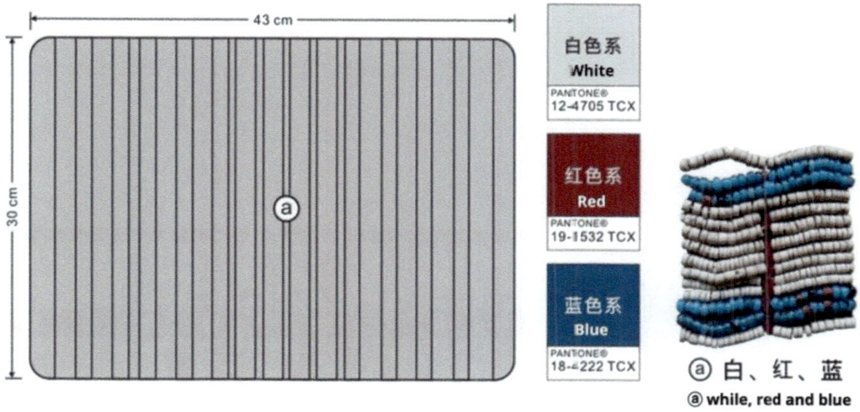

Fig. 4.21 Women's foot protection cloth (NO.54) (Drawn by this study)

of the Dajia River and its tributaries, consisting of seven tribes: S'ulay (Slightly Coming Tribe), Mepasing (Mopaxin, also known as the White Hair Tribe), Qsay' (A Leng e), South Qsya' (South A Leng), Kuras (Gulas, known as Jiulangxi from 1895 to 1945), Silaku (Xilaku), and Ulayruma (Wulailuma). The South Qsya split from the Qsya and settled in the mid-mountain area of A Leng Mountain on the right bank of the Beigang River in Nantou County [22].

Taking the NO.54 women's foot protection cloth as an example, it is made of a rectangular hemp cloth, using horizontal and vertical threading to connect shell beads and interweaving them on the cloth surface to form stripes. Material: ramie, shell beads. Techniques: plain weave (base), interweaving, bead weaving (shell beads). Pattern: stripes. Colors: white, red, blue. The fabric has no apparent pattern, mainly composed of stripes formed by connecting shell beads, as shown in Fig. 4.21ⓐ —the color combination of the beads: white, red, and blue. The positions of the patterns of other Nanshi textiles are shown in Fig. 4.22.

4.1.11 Meicyubus Textiles

From 1895 to 1945, it was under the jurisdiction of the Nanzhuang Sub-Hall of Hsinchu Hall (now located in Donghe Village, Nanzhuang Township, Miaoli County). It is in the mountainous area upstream of Nanzhuang River, northwest of Meicyubus Mountain, with only the Cyubus (Meicyubus) [23]. According to the description in the "Investigation of the Takasago: Overview of the Barbarian," the Meicyubus split from the Bailan of Zhudong County (belonging to the Tayal Jialarai) due to the need for farmland and moved to the southwestern foot of Dawo Mountain on the right bank of Bilin River. In 1902, due to the Nanzhuang Incident, they fled to the area around Meicyubus Mountain and then returned. In 1910, on the eve of the

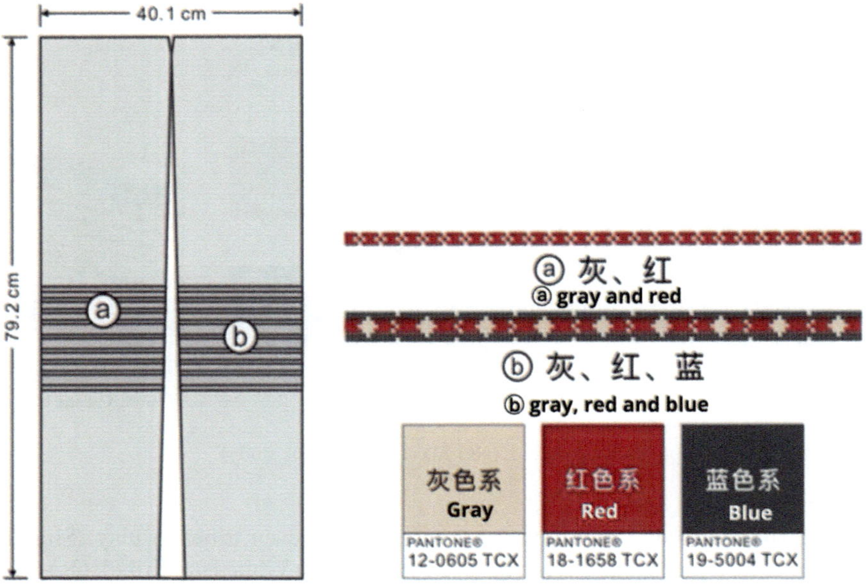

Fig. 4.22 Schematic of the pattern positions of the Nanshi textiles (Drawn by this study)

advance to the Narrow Brave Line, they fled outside the line and lived upstream of the North Pit River. In 1913, they were intimidated by the punitive expedition against the Xiake Luo and came to surrender. In September of that year, they moved to their current residence [24].

Taking the NO.55 men's sleeveless long robe as an example, this garment is relatively narrow and small, with woven patterns covering both the front and back. Below the center line on the back is a large diamond pattern, with shell bead decorations added to the sides and back seam. The whole piece has four types of composite stripes. Material: Ramie, shell beads. Techniques: Plain weave (base), warp heavy plain interweave, bead attachment. Patterns: Diamond pattern, zigzag pattern, × pattern, stripe pattern. Colors: Grey, red, blue. The composite stripes are arranged in a continuous square pattern, as shown in Figs. 4.23ⓐ ⓐ ⓑ ⓒ , with the diamond pattern as the main body, composed of smaller diamond structures. Figure 4.24ⓐ shows a large pattern composed of smaller diamonds with more diamonds than the former. The area of the woven pattern is also wider—a color combination of the pattern: Grey, red, and blue. The rest of the deer field group textiles and pattern position numbers are shown in Fig. 4.25.

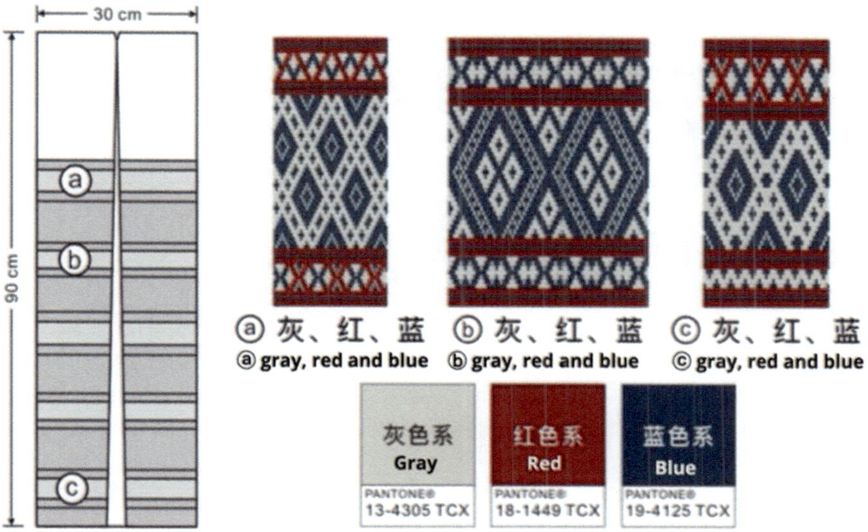

Fig. 4.23 Men's Sleeveless Long Robe—Front (NO.55) (Drawn by this study)

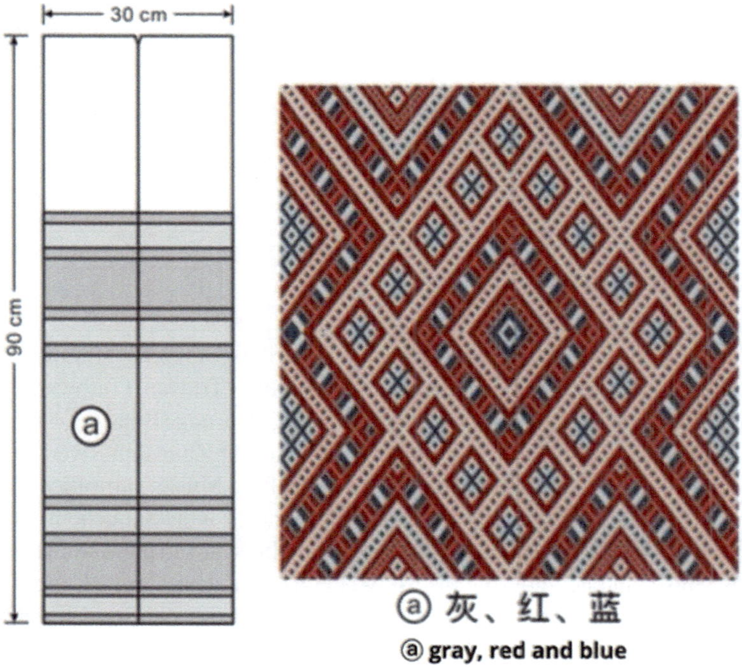

Fig. 4.24 Men's Sleeveless Long Robe—Back (NO.55)

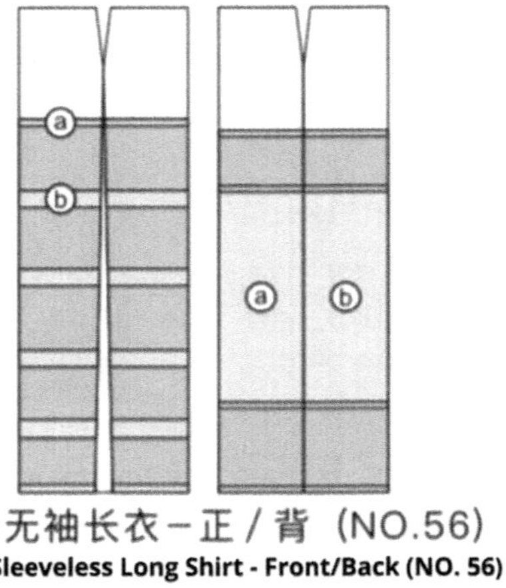

Fig. 4.25 Schematic of the pattern positions of the deer field group textiles (Drawn by this study)

4.2 Seediq

The Atayal Proper of the northwest and the Sədeq of the southeast are combined, commonly known as the Atayal tribe. The Sədeq and the eastern Təroko have become independent, respectively known as the Seediq and the Truku. Although the tribal names differ, the people of the two regions are consistent in language and customs [25]. This chapter focuses on the subgroups of the Seediq, describing their cultural background, geographical distribution, and textile characteristics, and presents the patterns and colors of Atayal textiles through mapping and comparison.

4.2.1 Musha Textiles

The Musha tribal system refers to the twelve tribes of Paran (Baran), Tongan (Tongan), Sipo, Tkanan (Takanan), Drodux (Rodov), Truwan (Tarowan), Gungu (Hogo), Mehebu (Mahopo), Qacuq (Kazuku), Suku, Boarung (Boarung) and Bkasan (Wukasan), which are scattered in the upper reaches of the Zhuoshui River, only three or four miles apart, and are under the jurisdiction of the Musha sub-office [26]. The territory of the Musha community extends from the north at the Shoucheng Mountain (Podian), to the south at the boundary between the Plnawan and the Bunun tribe, west to the Puli Basin and neighboring the Meiyuan group, and east to the Central Mountain Range. The tribes inhabiting the upper reaches of the Zhuoshui River are the Musha community (Tək-daya), the Toroko group (Toroko), and the Tuuda (Taudā), who share a common language and customs, maintaining an inseparable relationship. Among them, the Musha community holds a dominant position. Those who crossed the Central Mountain Range to the east and settled in the Pulibao Creek area of Hualien Hall became the Pulibau (Pulevao), who split from the Musha community.

Similarly, those who crossed the Central Mountain Range to the east and settled in the Liwu Creek, Sanzhan Creek, and Pulibao Creek areas of Hualien Hall formed the Truku (Təroko) and the Tausa (Tausa), who split from the Toroko and Tuudas in the source area of the Zhuoshui River in the west. The Truku tribe was originally a branch of the Atayal tribe that migrated eastward, belonging to the Seediq system. Old documents refer to them as "East Seediq" to distinguish them from the "West Seediq" centered around Musha in Ren'ai Township, Nantou County, in the west. The "West Seediq" has also become an independent tribe known as the Seediq. The two share the same lineage, and there is not much difference in their language and customs [27].

Taking the NO.61 men's and women's shawl as an example, it is made of three pieces of cloth. The red part is a compound diamond pattern, with cloth loops reserved for ties at the upper ends, and the whole piece has eight compound stripes. Material: ramie, cotton. Technique: plain weave (base), colored warp pick-up (double warp), longitudinal stripes. Pattern: diamond pattern, stripes. Color: black, red. The compound stripes are composed of diamond patterns, as shown in Fig. 4.26ⓐ ⓑ ⓒ ⓓ ⓔ ⓕ ⓖ ⓗ . These patterns are arranged horizontally in stripes by four to five diamond patterns, which seem to be arranged continuously in pairs, but the structures on the left and right sides of some single patterns are different, presenting different diamond patterns on the same compound stripe, as shown in Fig. 4.26ⓐ 1, 2, 3. Similar patterns are grouped, such as Fig. 4.26ⓐ 1, 4; ⓑ 1, 4; ⓒ 3; ⓓ 2; ⓔ 3; ⓕ 2, 3; ⓖ 2, 3, 4; ⓗ 2, 3, characterized by an internal structure composed of four small diamonds combined with an X shape. Other similar patterns, such as Fig. 4.26ⓐ 2, ⓒ 4, ⓓ 4, ⓔ 4 and Fig. 4.26ⓒ 1; ⓓ 1; ⓔ 1, 5; ⓕ 1; ⓗ 1, these patterns have very similar structures; the difference lies in the different directions of the four small diamond structures inside the pattern, creating variation within the same form. Figure 4.26ⓐ 3; ⓑ 2, 3; ⓕ 5; ⓖ 5; ⓗ 5 and Fig. 4.26ⓒ 2, ⓓ 3, ⓔ 2 and Fig. 4.26ⓕ 4, ⓗ 4 are all the same patterns. Figure 4.26ⓒ 5, ⓓ 5 should be the same pattern structure, but Fig. 4.26ⓒ 5 shows a different structural variation. Figure 4.26ⓖ 1 is the only one without any similar structure. This piece of fabric mainly has eight pattern structures, among which it is worth noting that Fig. 4.26ⓒ 5, ⓓ 5, ⓔ 5, ⓕ 2, ⓖ 4 structures seem to have errors or asymmetrical weaves deliberately. The color combination of the patterns is black and red. The positions and numbers of other patterns and textiles of the Musha community are shown in Fig. 4.27.

4.2.2 Pulibau Textiles

The Pulibau, named after their residence on both sides of the Pulibao Creek, originated from the Chunyang Hot Springs area in the upper reaches of the Zhuoshui Creek in Nantou and the upper reaches of the Mei Creek. Some of the tribe members, in search of hunting grounds and farmland, crossed the Central Mountain Range and entered the upper reaches of the Pulibao Creek in Hualien, residing in the mountainous area between today's Longjian and Wenlan [28]. The tribal groups in the

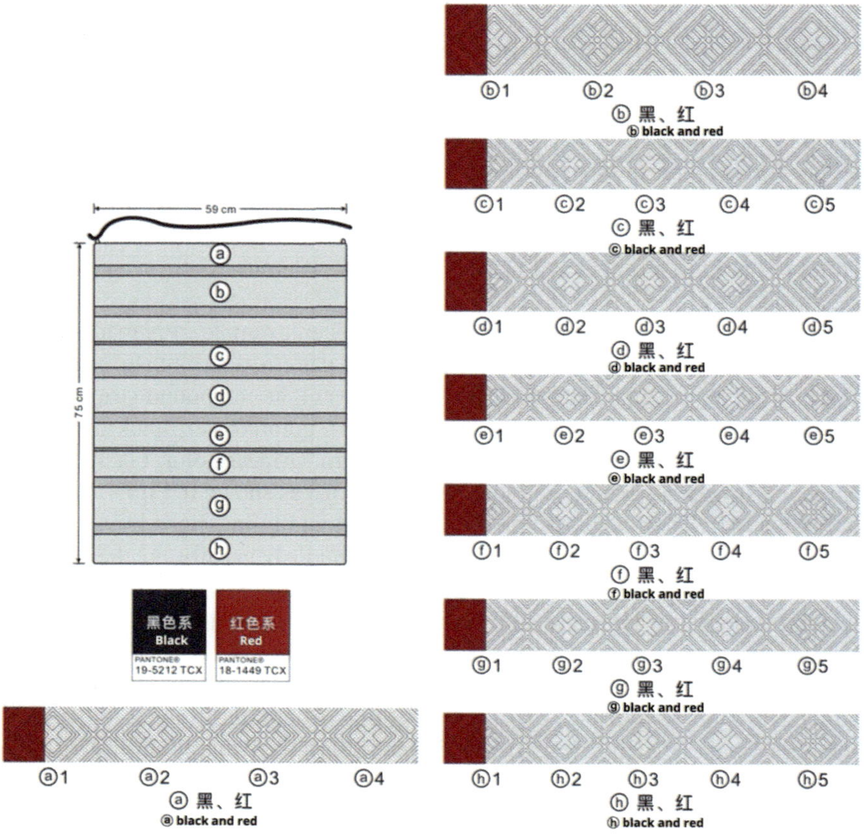

Fig. 4.26 Men's and Women's Shawl (NO.61) (Drawn by this study)

Pulibao Creek basin do not have their tribal names and are referred to as Pulevao by the Truku and Tsungao by the Amis tribe. The Pulibau once held its ground on both sides of the Pulibao Creek, standing toe-to-toe with the powerful Truku group in the north, and frequently attacked the Bunun Dan group and various Amis tribes in the south and east, resulting in a significant cause of the Amis tribe's migration. The Pulibau claims to be a branch separated from Tək-daya-Tərowan, the Truku Bay group of the Musha. The Pulibau still retain the language and customs of their ancestral Musha. For example, the Pulibau's practice of tattooing several vertical lines on their foreheads is a notable example. Especially the characteristic facial tattoos of the Pulibau, which the Amis tribe uses as a basis to identify the Truku or Pulibau [29].

Taking NO.63 men's sleeveless long robe as an example, the robe comprises two pieces of hemp fabric joined together, with tassel decoration at the center line of the back. The entire piece is composed of several diamond patterns interwoven, with patterns on both the front and back. This piece of textile is slightly

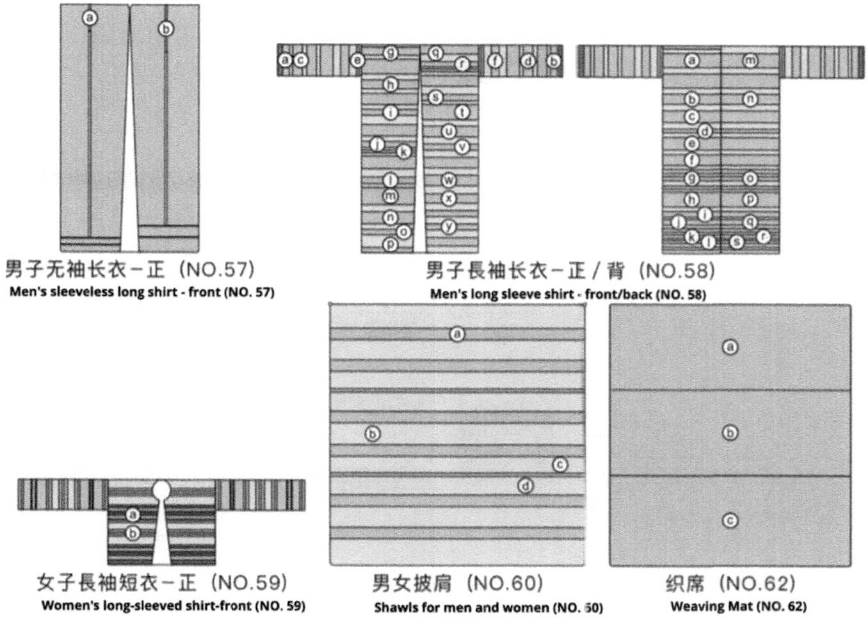

Fig. 4.27 Schematic of the pattern placement on Musha textiles (drawn by the author of this study)

damaged, but the pattern can still be identified, with 12 kinds of compound stripes in total. Material: ramie, wool. Techniques: plain weave (base), colored weft floating weave, colored weft pick-up inlay. Patterns: diamond pattern, mortar pattern, mountain pattern, stripe. Colors: black, grey, white, red. Compound stripes are arranged in pairs, as shown in Figs. 4.28ⓐ ⓑ ⓒ ⓓ ⓔ and 4.29ⓐ ⓑ ⓒ ⓓ ⓔ ⓕ ⓖ . Figures 4.28ⓑ ⓒ show the same compound stripes, with diamond patterns interwoven with mortar patterns within the stripes. Similar patterns are grouped, such as Figs. 4.28ⓑ ⓒ , 4.28ⓓ ⓔ , 4.29ⓓ , 4.29ⓒ ⓕ , and 4.29ⓔ ⓖ , which are all structurally similar. Figure 4.29ⓑ has a more extensive and more straightforward pattern structure. Pattern color combinations: black, red; black, white; white, red.

4.2.3 Truku Textiles

The Truku people are distributed in the mountainous areas east of the Central Mountain Range, starting from the right bank of the Pulibao Creek in the south to the right bank of the Da Zhuo Shui River in the north. They are further divided into three parts: the Inner Truku (originally residing in the Pingsheng tribe of Ren'ai Township, Nantou County, they crossed the Central Mountain Range in the seventeenth century and moved east to the upper and middle reaches of the Liwu River in Xiulin Township, known as the Truku), the Outer Truku (expanded from the Inner Truku

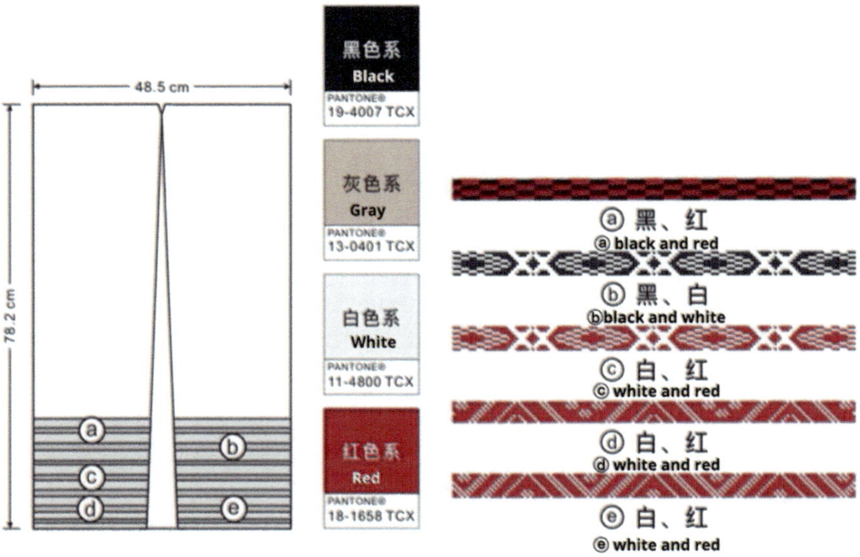

Fig. 4.28 Men's Sleeveless Long Robe—Front (NO.63) (drawn by the author of this study)

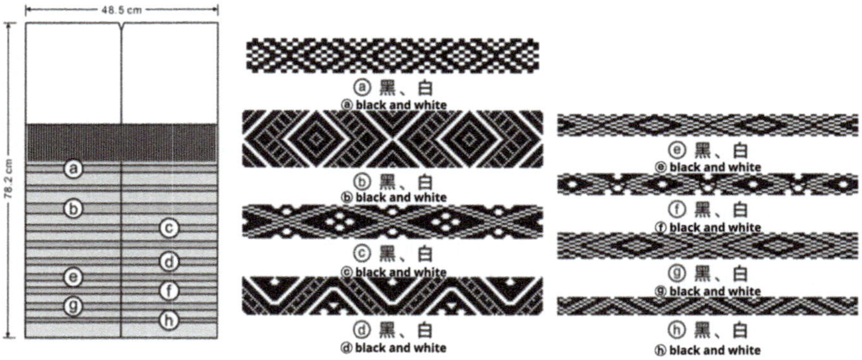

Fig. 4.29 Men's Sleeveless Long Robe—Back (NO.63) (drawn by the author of this study)

to the east coast, reaching the lower reaches of the Liwu River, Sanzhan River and Heping River in Xiulin Township, also known as the Truku), and the Mijolan (now known as Batolan) [30].

Taking the NO.69 women's long skirt as an example, it is made of three pieces of cloth, edged with red wool, with cloth rings on both sides of the upper end, and has two patterns. Material: ramie, wool, cotton. Technique: plain weave (base), color discontinuous weft inlay. Pattern: diamond pattern. Colors: black, white, red, peach red, green, blue. The pattern is a variation of the diamond pattern, arranged in columns of six identical patterns, with spaces left between the patterns, and the

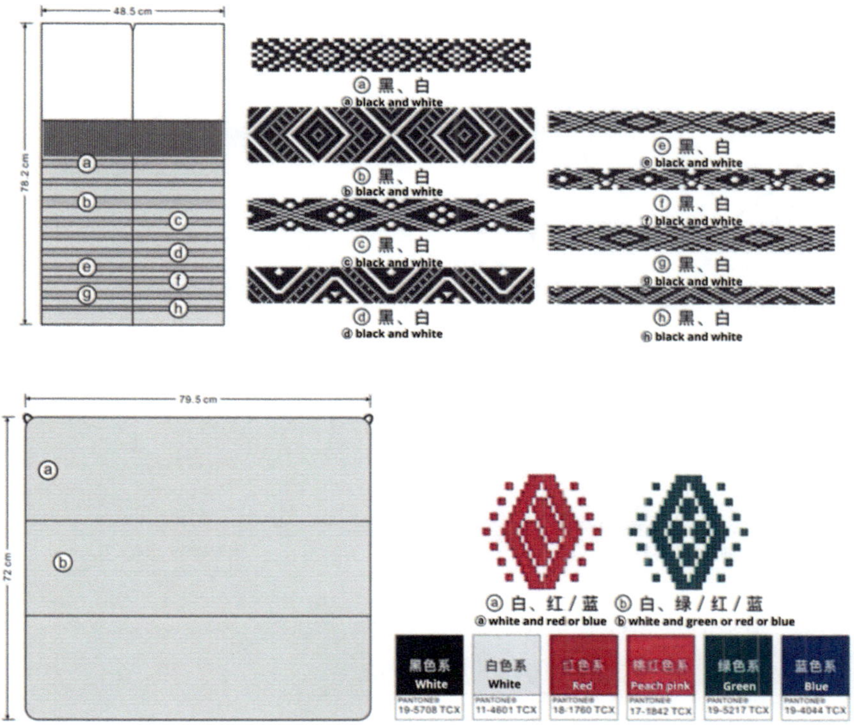

Fig. 4.30 Women's long skirt (NO.69) (Drawn by this study)

patterns in two straight columns are staggered, as shown in Fig. 4.30ⓐ ⓑ. These patterns look very similar; the characteristic is that there are different structures in the middle of the pattern, such as two diagonal lines and four crosses as the core of a single pattern—color combinations of the pattern: white, red, white, green. The same pattern will use different colors, such as the color combination of white and red. Red will be replaced by blue, so white and green (red and blue) will also match this way. The rest of Truku's textiles and pattern position numbers are shown in Fig. 4.31.

4.3 Textiles of Unknown Subgroups

The traditional costumes of the minority ethnic groups in Taiwan vary greatly and can generally be identified by the patterns and colors of their clothing. The Atayal tribe is the most widely distributed ethnic group among the minorities in Taiwan, and the characteristics of their traditional costumes are distinct, making it easy to distinguish them from other tribes. However, to further categorize the Atayal textiles into subgroups or locations, one must rely on descriptions of the original collection sites, compare them with the original collection records of the same textiles owned

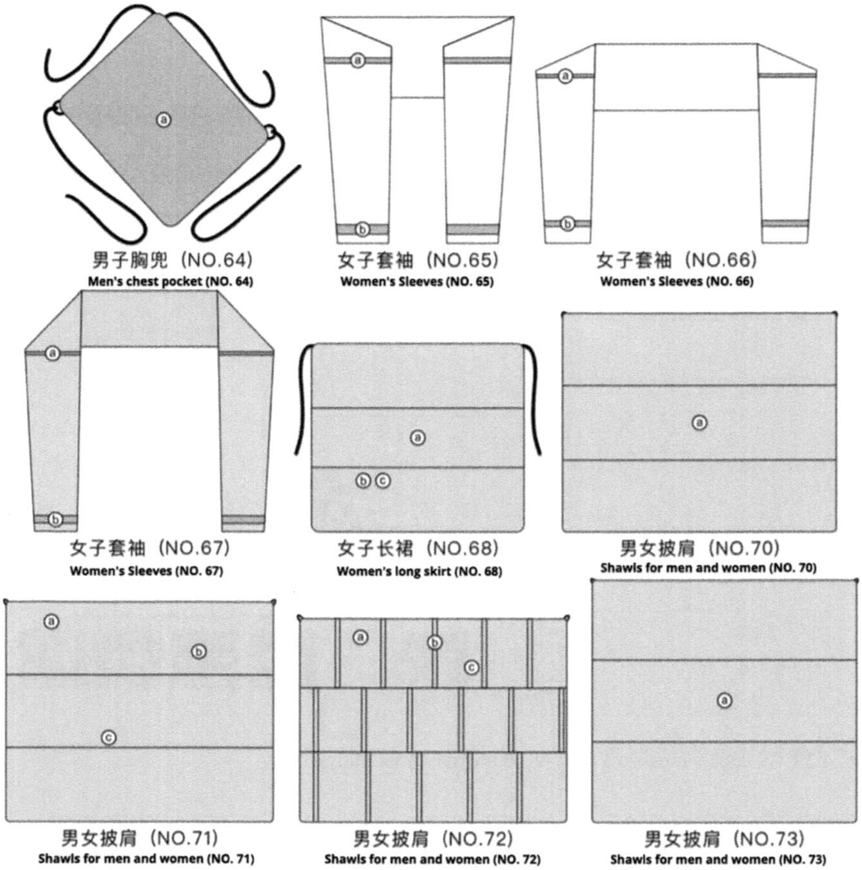

Fig. 4.31 Schematic of the pattern positions of the Truku textiles (Drawn by this study)

by other museums, and corroborate with relevant historical documents. This is one of the methods of categorizing the traditional costumes of minority ethnic groups. If the above information cannot be confirmed or is untraceable, comparison through the characteristics of cultural relics is also an important clue to infer the subgroup to which the textile belongs. Due to the wide distribution area of the Atayal tribe and the migration of subgroups, there may be many uncertainties when comparing patterns and colors. In this study, we often encounter Atayal textiles that cannot be categorized into subgroups when selecting research samples. Although these are excluded from the research samples, it illustrates the difficulty of investigating cultural relics during the research process. Here, we specifically cite the two traditional costumes that cannot be categorized into subgroups, as shown in Figs. 4.32, 4.33, and 4.34, as examples and case presentations.

Take NO.74 men's sleeveless long robe as an example; the body of the robe is made up of two pieces of linen joined together, with patterns concentrated on the

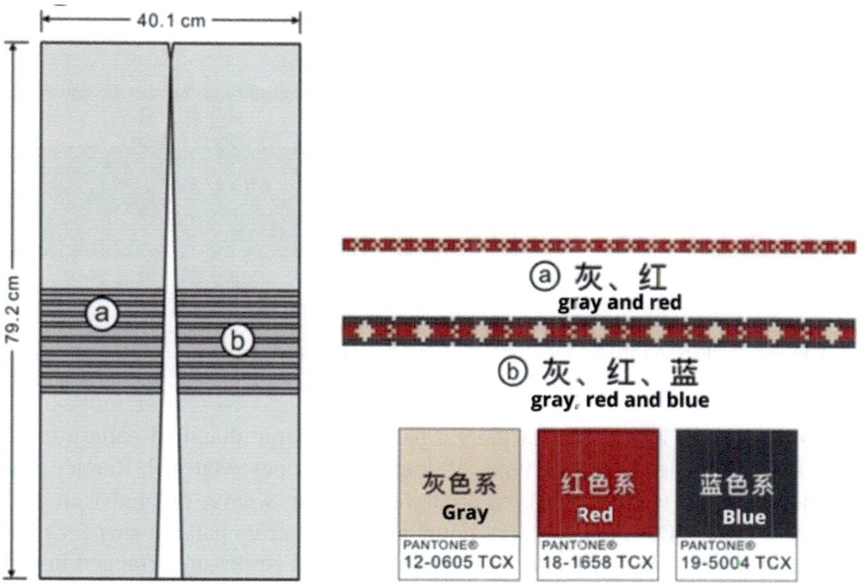

Fig. 4.32 Men's Sleeveless Long Robe—Front (NO.74) (Drawn by this study)

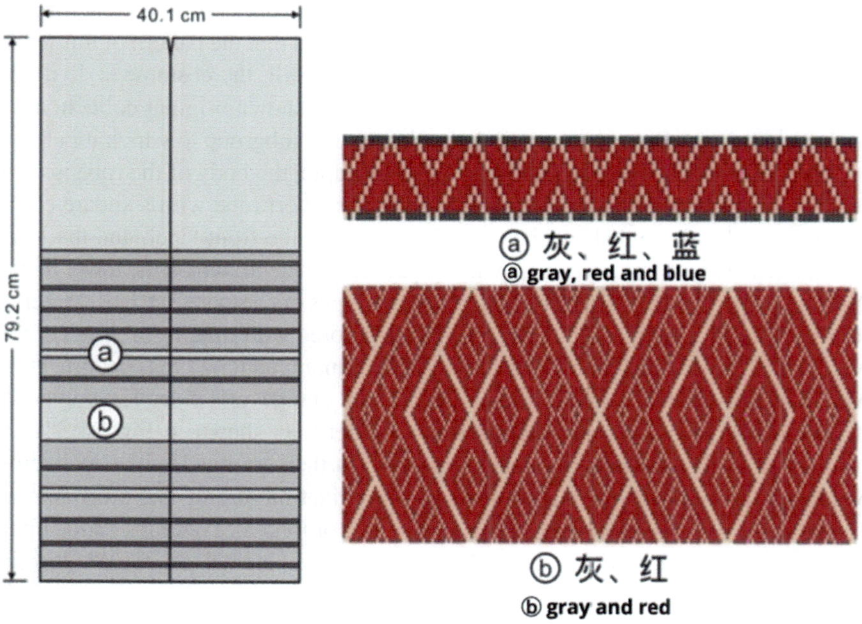

Fig. 4.33 Men's Sleeveless Long Robe—Back (NO.74) (Drawn by this study)

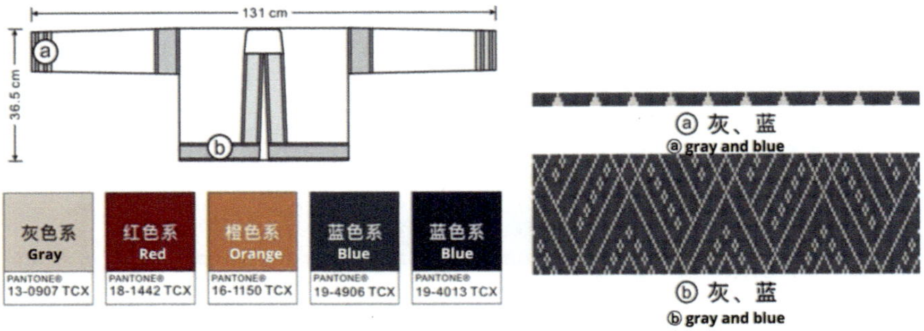

Fig. 4.34 Long-sleeved robe—Front (NO.75) (Drawn by this study)

front, middle, and lower back of the garment and a large diamond pattern on the back. The whole piece has four types of compound stripes. Material: Ramie, wool. Techniques: Plain weave (base), colored weft floating weave, colored weft pick-up inlay. Patterns: Diamond pattern, mountain pattern, cross pattern, stripe. Colors: Grey (natural color of linen), red, blue. The compound stripes are arranged in pairs continuously, as shown in Figs. 4.32ⓐ ⓑ and 4.33ⓐ ⓑ . Figures 4.32ⓐ ⓑ show small patterns, and Figs. 4.33ⓐ ⓑ show large patterns, in which the × structure inside the diamond pattern in Fig. 4.33ⓑ comprises four identical small diamonds. Pattern color combinations: Grey, red; Grey, red, blue. This sleeveless long robe is a typical style of Atayal men's clothing. This study found that the pattern of this textile is like the style of Klesan Township, Yilan County. Still, the costume style differs from the Klesan's traditional costume. Due to the unknown original collection site of the textile, it is currently impossible to identify the subgroup to which it belongs.

Taking the NO.75 long-sleeved robe as an example, the body of the robe is made up of two pieces of linen joined together to form a short robe with a square collar, symmetrical lapels, and sleeve buttons. Both sleeves, the front lapel, and the lower hem are woven with colored wool, with the patterns concentrated on the lower hem at the front of the garment. The whole piece has two types of composite stripes. Material: Ramie, wool. Techniques: Plain weave (base), colored weft floating weave, colored weft picked flower sandwich weave. Patterns: Mountain pattern, cross pattern, stripe pattern. Colors: Grey (original color of linen), red, orange, and blue. The composite stripes are arranged in a continuous square pattern, as shown in Figs. 4.34ⓐ ⓑ . Figure 4.34ⓑ comprises mountain patterns, further arranged with small cross patterns to form a diagonal pattern—a color combination of grey and blue patterns. The Atayal tribe wears sleeveless long tops (vests) or long-sleeved tops on the upper body, with a bra worn underneath. According to the original collector's description of the museum collection, this garment may also be the traditional attire of the Paiwan or Saisiyat. This study verifies that the pattern of this textile is like the style of Biho Village in Klesan Township, Yilan County. Still, the style of the garment is vastly different from the traditional attire of the Klesan. Due to the unknown original collection location of the textile, it is currently impossible to identify the sub-group it belongs to.

4.4 Museum Collection of Atayal Textiles

Exploring the background origins of the Atayal sub-groups and mapping and comparing the patterns and colors of Atayal textiles provides a further understanding of their clothing culture. To understand the relationship between the artifact specimens and their respective sub-groups, this study refers to "The Atayal People of Klesan: Ethnographic Field Research and Study Volume I and II" [31], "Survey Report on Aboriginal Customs: Volume I Atayal" [32], "Survey Report on Aboriginals: Volume VII Atayal Part II, Volume IV Seediq and Truku, Volume V Atayal Part I" [33], "Survey Book on the Takasago Tribe: Overview of Aboriginal Societies" [34], "Research on the System of Taiwan's Minority Ethnic Groups: Volume I Main Text, Volume II Data" [35], "Comprehensive Dictionary of Taiwan's Minority Ethnic Groups' History, Language, and Culture".[3] The literature provides an understanding of the background and origins of the subgroups. The survey content from 1895 to 1945 mainly focuses on individual names, family names, and surnames within the ethnic group (subgroup), narrating to clans and tribes and then discussing the general outline of tribal composition. Based on the distribution of racial appearances, their places of origin, and old tribes, the paths of ethnic group migration and dispersion are inferred. The survey results generally also target the origins, migrations, and settlements of the subgroups, with similar descriptions, such as the three major language group systems of the Atayal tribe's Seediq, Truku, and Tayal, all belong to the high mountain origin theory. Due to small arable land, population increase, lack of resources, and hostile relations, some people moved out to seek new places to live, exploring new farmland and hunting grounds. New residences, such as the Malepa and Mabarnor of the Atayal tribe, are usually chosen near rivers and valleys to meet living needs. Of course, the migration process will likely have more contact and interaction with different ethnic groups (or other Atayal subgroups), naturally leading to tribes mixed with various ethnic groups or creating new groups.

This study has a total of 73 research samples. The subgroups they belong to can understand the relationship between each other from the origin of myths and legends, as well as the migration process and dialect language system, such as the Hakul (Hakku) originating from the Binspogan rift rock, migrating out of the Makanji, Malepa, Taranan and some of the Nanshi, come from the Malepa. The subgroup coming from the Damjian Mountain, such as the Nanshi, is a collection of many molecules, which can be divided into the Seediq system and the Slightly, Baimao, and Aleng communities that seem to belong to the Beishi. The ancestors originated from the boulder of Baishishan, such as the Pulibau split from the Musha, which is proven to be the result of the early Atayal tribe's migration to the east. The examples above of subgroup migration are just a tiny part of the Atayal subgroups. Still, they may be connected to the large group and then explore the relationship between textile patterns and subgroups.

[3] Taiwan Minority Ethnic Groups History, Language, and Culture Dictionary, https://reurl.cc/6Nky6O, 3rd May 2023.

The traditional costumes of the Atayal tribe have a significant ethnic style, and the costumes can also be a symbol to distinguish the ethnic group. For example, the Atayal people live in multi-ethnic areas, and there are other ethnic groups, such as the Saisiyat, Bunun, and Amis, nearby. These ethnic groups have long been in the same region and have communicated with each other due to their close living circles. Clothing has become an essential symbol of basic ethnic recognition and separation from different ethnic groups. Of course, the various subgroups of the Atayal tribe also use this as a recognition method. Analyzing the cultural representation of clothing and visually recognizing its physical properties, that is, style, material, technique, pattern, color, etc., all belong to the cultural structure and have apparent symbolic features. The traditional costumes of the Atayal tribe are mostly homemade hemp spun yarn, which is tailored after weaving. Cotton, wool, and other yarn materials were mixed in the later period. In terms of weaving techniques, they are best at weaving. Weaving and weaving are entirely different techniques, but they are ways to organize soft yarn into structures. The methods used include plain weave, twill weave, floating weave, picking weave, sandwich weave, braiding, etc. The textile patterns are generally visible, including diamond patterns, mountain patterns, zigzag patterns, cross patterns, X-shaped patterns, etc., as form changes, forming composite stripes or single patterns on the clothing.

The patterns and colors of Atayal tribal textiles are diverse, but they exhibit similar woven structures due to the limitations of weaving tools and techniques. Some subgroups' traditional costumes may influence each other due to their geographical proximity, presenting the same style, such as NO.34 Bunsui (men's and women's long-sleeved robes), NO.35 Bunsui (men's long-sleeved robes), NO.36 Bunsui (men's long-sleeved robes), NO.39 Bunsui (women's long-sleeved robes), NO.40 Bunsui (women's long-sleeved robes), and NO.41 Beishi (men's long-sleeved robes), NO.42 Beishi (men's long-sleeved robes), NO.44 Beishi (women's long-sleeved robes), NO.45 Beishi (women's long-sleeved robes), NO.46 Beishi (women's long-sleeved short robes), NO.47 Beishi (women's long-sleeved robes) etc. All these textiles have black cotton fabric sewn at the cuff position, as shown in Figs. 4.35 and 4.36. This feature may be designed to prevent the cuffs from getting dirty quickly. Looking at the textiles of different Atayal subgroups, they can create many different patterns and color combinations within the unified tribal style, undoubtedly providing another design thinking that can break through the standard norms under conditions, fully demonstrating the phenomenon of cultural re-creation of the tribe.

The compound stripes of Atayal tribal textiles can be composed of different patterns, an essential feature of Atayal tribal textiles. When depicting patterns in this study, it was found that the internal structure of compound stripes is only sometimes symmetrical up and down or left and right. Some are arranged in asymmetrical expressions with different patterns. This phenomenon should be a deliberately created woven effect, such as NO.01 Hakul (Hakku) (men's and women's shawls), NO.04 Msbtunux (men's and women's shawls), NO.05 Klesan (men's sleeveless robes), NO.06 Klesan (men's sleeveless robes), NO.17 Klesan (women's belts), NO.19 Klesan (women's belts), NO.36 Bunsui (men's long-sleeved robes), NO.56 Meicyubus (men's sleeveless robes), NO.61 Musha (men's and women's shawls),

Fig. 4.35 Men's and Women's Long-Sleeved Robes—Front (NO.34) (Photographed by this study)

Fig. 4.36 Men's Long-Sleeved Robes—Front (NO.41) (Photographed by this study)

NO.62 Musha (woven mats), NO.63 Pulibau (men's sleeveless robes) etc., all can perceive the asymmetrical structure of compound stripes. Although asymmetrical patterns are created within the compound stripes, they do not develop a sense of disorder due to the excellent harmony of color matching.

The collection of Atayal textiles is like a scattered history book. Although it is possible to collate the threads and relationships of the textile group through various verifications, it is still within a limited range, striving to interpret the background information of the cultural relics' specimens and gradually outline the style of Atayal textiles. In recent years, the attention paid to ethnic minorities in Taiwan has surpassed the past. From official to museums and other related institutions, they try to convey the unique ethnic culture of ethnic minorities through various activities. The purpose of establishing a museum as part of the collection of ethnic minority cultural relics can be said to have completed the stage work, and the objects that can be collected are limited. Instead, the work of display, education, and research continues through different channels, allowing the general public to directly or indirectly understand the culture of Taiwan's ethnic minorities, such as the permanent exhibitions of ethnic minority cultural relics in the Taiwan Museum, National Taiwan University Museum of Anthropology, Taiwan Prehistoric Culture Museum, and the special exhibition of ancient Paiwan cultural relics returning to Lion Township, Pingtung County in 2013, and the annual Taiwan Ethnic Minority Research Paper Conference, all based on the concrete actions of inheriting the culture of Taiwan's ethnic minorities. However, in studying the cultural relics of ethnic minorities in the museum, in addition to discovering new material cultural clues, trying to propose different views, and looking for a framework to explain ethnic art, it is also expected to convey the "beauty of ethnic groups" through the education mechanism.

References

1. Walis Nokan, Yu Guanghong: "*History of Taiwan's Minority Ethnic Groups: Atayal History (台湾少数民族历史: 泰雅族史篇)*," Taiwan Literature Museum, 2002 edition, page 1.
2. Folklore and Anthropology Research Room, Taipei University: "*Research on the System of Taiwan's Minority Ethnic Groups: Volume One Main Text (台湾少数民族系统所属之研究: 第一册本文篇)*," translated by Yang Nanjun, Taiwan Minority Ethnic Committee, Nantian Bookstore Co., Ltd., Taiwan, 2011 edition, page 23.
3. Taiwan Old Custom Survey Association: "*Aboriginal Custom Survey Report: Volume One, Atayal Tribe (番族惯习调查报告书: 第一卷泰雅族)*," translated by the Institute of Ethnology, Academia Sinica, Taiwan, 1996 edition, page 17.
4. Taiwan Old Custom Survey Association: "*Aboriginal Custom Survey Report: Volume Five, Atayal Tribe Part One (蕃族调查报告书: 第五册泰雅族前篇)*," translated by the Institute of Ethnology, Academia Sinica, Taiwan, 2012 edition, page 21.
5. Taiwan Old Custom Survey Association: "*Aboriginal Custom Survey Report: Volume One, Atayal Tribe (番族惯习调查报告书: 第一卷泰雅族)*," translated by the Institute of Ethnology, Academia Sinica, Taiwan, 1996 edition, pages 14–15.
6. Taiwan Old Custom Survey Association: "*Aboriginal Custom Survey Report: Volume One, Atayal Tribe (番族惯习调查报告书: 第一卷泰雅族)*," translated by the Institute of Ethnology, Academia Sinica, Taiwan, 1996 edition, page 19.
7. Taiwan Old Custom Survey Association: "*Aboriginal Custom Survey Report: Volume Five, Atayal Tribe Part One (蕃族调查报告书: 第五册泰雅族前篇)*," translated by the Institute of Ethnology, Academia Sinica, Taiwan, 2012 edition, page 20.
8. Taipei University Folklore and Anthropology Research Room: "*Research on the System of Taiwan's Minority Ethnic Groups: Volume One, Main Text (台湾少数民族系统所属之研究: 第一册本文篇)*," translated by Yang Nanjun, Taiwan Minority Ethnic Committee, South Sky Bookstore Co., Ltd., Taiwan, 2011 edition, pages 32–34.
9. Li Yiyuan, Shi Lei, Ruan Changrui, Yang Fufa: "*The Atayal People of Klesan: Ethnological Field Research and Study Volume One (南澳的泰雅人: 民族学田野调查与研究上册)*," Institute of Ethnology, Academia Sinica, Taiwan, 1963 edition, page 8.
10. Taiwan Old Customs Survey Association: "*Aboriginal Investigation Report: Volume 5, Atayal Tribe Part 1 (蕃族调查报告书: 第五册泰雅族前篇)*," translated by the Institute of Ethnology, Academia Sinica, Taiwan, 2012 edition, page 22.
11. Taipei University Folklore and Anthropology Research Department: "*Research on the System of Taiwan's Minority Ethnic Groups: Volume 2, Data Volume (台湾少数民族系统所属之研究: 第二册资料篇)*," translated by Yang Nanjun, Taiwan Minority Ethnic Groups Committee, Taiwan Nantian Bookstore Co., Ltd., 2012 edition, page 27.
12. Taiwan Old Customs Survey Association: "*Aboriginal Investigation Report: Volume 7, Atayal Tribe Part 2 (蕃族调查报告书: 第七册泰雅族后篇)*," translated by the Institute of Ethnology, Academia Sinica, Taiwan, 2010 edition, page 8.
13. Taiwan Old Customs Survey Association: "*Aboriginal Custom Investigation Report: Volume 1, Atayal Tribe (番族惯习调查报告书: 第一卷泰雅族)*," translated by the Institute of Ethnology, Academia Sinica, Taiwan, 1996 edition, page 19.
14. Taiwan Old Customs Survey Association: "*Aboriginal Investigation Report: Volume 5, Atayal Tribe Part 1 (蕃族调查报告书: 第五册泰雅族前篇)*," translated by the Institute of Ethnology, Academia Sinica, Taiwan, 2012 edition, page 22.
15. Taipei University Folklore and Anthropology Research Department: "*Research on the System of Taiwan's Minority Ethnic Groups: Volume 2, Data Volume (台湾少数民族系统所属之研

究: 第二册资料篇)," translated by Yang Nanjun, Taiwan Minority Ethnic Groups Committee, Taiwan Nantian Bookstore Co., Ltd., 2012 edition, page 83.
16. Taiwan Old Customs Survey Association: "*Aboriginal Custom Investigation Report: Volume 1, Atayal Tribe (番族惯习调查报告书: 第一卷泰雅族),*" translated by the Institute of Ethnology, Academia Sinica, Taiwan, 1996 edition, pages 15-169.
17. Taiwan Old Customs Survey Association: "*Aboriginal Investigation Report: Volume 7, Atayal Tribe Part 2 (蕃族调查报告书: 第七册泰雅族后篇),*" translated by the Institute of Ethnology, Academia Sinica, Taiwan, 2010 edition, page 6.
18. Taipei University Folklore and Anthropology Research Department: "*Research on the System of Taiwan's Minority Ethnic Groups: Volume 1, Main Text Volume (台湾少数民族系统所属之研究: 第一册本文篇),*" translated by Yang Nanjun, Taiwan Minority Ethnic Groups Committee, Taiwan Nantian Bookstore Co., Ltd., 2011 edition, pages 73–74.
19. Taiwan Old Custom Survey Association: "*Aboriginal Investigation Report: Volume 5, Atayal Tribe Part 1 (蕃族调查报告书: 第五册泰雅族前篇),*" translated by the Institute of Ethnology, Academia Sinica, Taiwan, 2012 edition, page 21.
20. Taiwan Old Custom Survey Association: "*Aboriginal Custom Investigation Report: Volume 1, Atayal Tribe (番族惯习调查报告书: 第一卷泰雅族),*" translated by the Institute of Ethnology, Academia Sinica, Taiwan, 1996 edition, page 16.
21. Folklore and Anthropology Research Room, Taipei University: "*Research on the Affiliation of Taiwan's Minority Ethnic Groups: Volume 1, Main Text (台湾少数民族系统所属之研究: 第一册本文篇),*" translated by Yang Nanjun, Minority Ethnic Committee of Taiwan, South Sky Bookstore Ltd., Taiwan, 2011 edition, pages 79–80.
22. Taiwan Old Custom Survey Association: "*Aboriginal Investigation Report: Volume 5, Atayal Tribe Part 1(蕃族调查报告书: 第五册泰雅族前篇),*" translated by the Institute of Ethnology, Academia Sinica, Taiwan, 2012 edition, page 21.
23. Taiwan Old Custom Survey Association: "*Aboriginal Investigation Report: Volume 7, Atayal Tribe Part 2 (蕃族调查报告书: 第七册泰雅族后篇),*" translated by the Institute of Ethnology, Academia Sinica, Taiwan, 2010 edition, page 6.
24. Police Affairs Bureau, Aboriginal Affairs Section: "*Investigation Report on the Takasago Tribe: Aboriginal Community Overview (高砂族调查书: 蕃社概况),*" translated by the Institute of Ethnology, Academia Sinica, Taiwan, 2011 edition, page 107.
25. Folklore and Anthropology Research Room, Taipei University: "*Research on the Affiliation of Taiwan's Minority Ethnic Groups: Volume 1, Main Text (台湾少数民族系统所属之研究: 第一册本文篇),*" translated by Yang Nanjun, Minority Ethnic Committee of the Executive Yuan, Taiwan, South Sky Bookstore Ltd., Taiwan, 2011 edition, page 24.
26. Taiwan Old Custom Survey Association: "*Aboriginal Investigation Report: Volume 4, Seediq Tribe and Truku Tribe (蕃族调查报告书: 第四册赛德克族与太鲁阁族),*" translated by the Institute of Ethnology, Academia Sinica, Taiwan, 2011 edition, pages 13–14.
27. Folklore and Anthropology Research Room, Taipei University: "*Research on the Affiliation of Taiwan's Minority Ethnic Groups: Volume 1, Main Text (台湾少数民族系统所属之研究: 第一册本文篇),*" translated by Yang Nanjun, Minority Ethnic Committee of Taiwan, South Sky Bookstore Ltd., Taiwan, 2011 edition, pages 86–88.
28. Taiwan Old Customs Survey Association: "*Aboriginal Investigation Report: Volume 4, Seediq and Truku Tribes (蕃族调查报告书: 第四册赛德克族与太鲁阁族),*" translated by the Institute of Ethnology, Academia Sinica, Taiwan, 2011 edition, page 2.
29. Folklore and Anthropology Research Room, Taipei University: "*Research on the Affiliation of Taiwan's Minority Ethnic Groups: Volume 1, Main Text (台湾少数民族系统所属之研究: 第一册本文篇),*" translated by Yang Nanjun, Taiwan Minority Ethnic Groups Committee, Nantian Bookstore Co., Ltd., Taiwan, 2011 edition, page 108.
30. Taiwan Old Customs Survey Association: "*Aboriginal Investigation Report: Volume 4, Seediq and Truku Tribes (蕃族调查报告书: 第四册赛德克族与太鲁阁族),*" translated by the Institute of Ethnology, Academia Sinica, Taiwan, 2011 edition, page 1.
31. Li Yiyuan, Shi Lei, Ruan Changrui, Yang Fufa: "*The Atayal People of Klesan: Ethnological Field Investigation and Research, Volumes 1 and 2 (南澳的泰雅人: 民族学田野调查与研究上册、下册),*" Institute of Ethnology, Academia Sinica, Taiwan, 1963 and 1964 editions.

32. Taiwan Old Customs Survey Association: "*Aboriginal Custom Investigation Report: Volume 1, Atayal Tribe (番族惯习调查报告书: 第一卷泰雅族),*" translated by the Institute of Ethnology, Academia Sinica, Taiwan, 1996 edition.
33. Taiwan Old Customs Survey Association: "*Aboriginal Investigation Report: Volume 7, Atayal Tribe Part 2, Volume 4, Seediq and Truku Tribes, Volume 5, Atayal Tribe Part 1 (蕃族调查报告书: 第七册泰雅族后篇、第四册赛德克族与太鲁阁族、第五册泰雅族前篇),*" translated by the Institute of Ethnology, Academia Sinica, Taiwan, 2010, 2011, and 2012 editions.
34. Police Affairs Bureau, Aboriginal Affairs Section: "*Investigation of the Takasago Tribe: Aboriginal Community Overview (高砂族调查书: 蕃社概况),*" translated by the Institute of Ethnology, Academia Sinica, Taiwan, 2011 edition.
35. Folklore and Anthropology Research Room, Taipei University: "*Research on the Affiliation of Taiwan's Minority Ethnic Groups: Volume 1, Main Text, Volume 2, Data Volume (台湾少数民族系统所属之研究: 第一册本文篇、第二册资料篇),*" translated by Yang Nanjun, Taiwan Minority Ethnic Groups Committee, Nantian Bookstore Co., Ltd., Taiwan, 2011 and 2012 editions.

Chapter 5
Characteristics of Pan-Atayal Textile Patterns

5.1 Characteristics of Textile Pattern Shapes

This study analyses and summarizes the clothing and pattern types of 73 Atayal textiles. Men's clothing can be divided into sleeveless long clothes, sleeveless short clothes, long-sleeved long clothes, long-sleeved short clothes, breastplates, and open-back trousers, totaling 35 pieces. Women's clothing can be divided into long-sleeved long clothes, long-sleeved short clothes, sleeve sets, belts, long skirts, and foot protection clothes, totaling 21 pieces. Textiles that can be worn or used by both men and women, such as long-sleeved long clothes (NO.34), tobacco bags, woven mats, and a more significant number of shawls, totaling 17 pieces. These textile patterns include variable patterns, diamond patterns, circular patterns, mortar patterns, zigzag patterns, square patterns, mountain patterns, cross patterns, X-shaped patterns, netted stripes, dotted stripes, and stripes, totaling 12 types of pattern types, as shown in Appendix 2. Among them, diamond patterns (29.6%), stripes (23.9%), mountain patterns (15%), cross patterns (7.5%), zigzag patterns (7%), X-shaped patterns (5.6%), and circular patterns (3.8%) appear the most, followed by square patterns (2.8%), mortar patterns (1.9%), dotted stripes (1.4%), netted stripes (0.9%), and variable patterns (0.5%).

In the 73 research samples, every type of textile has at least one pattern, such as NO.25, NO.38, NO.52, NO.64 men's chest pockets; NO.27 men's sleeveless long coats; NO.28, NO.68, NO.69 women's long skirts; NO.54 women's foot protection cloth; NO.59 women's long-sleeved short coat; NO.62 woven mat; NO.70, NO.71, NO.73 men's and women's shawls. The same piece of clothing can have up to 8 patterns, such as NO.05 men's sleeveless long coat with diamond pattern, circular pattern, zigzag pattern, square pattern, mountain pattern, cross pattern, X pattern, stripe pattern, etc. Looking at the types and numbers of these textile patterns, there are 14 pieces with only one pattern, 18 pieces with two patterns, 17 pieces with three patterns, 14 pieces with four patterns, seven pieces with five patterns, two pieces with six patterns, only one piece with eight patterns, among which the single pieces with

P. H. Wang and J. Li, *Interpreting the Evolution of Atayal Tribal Textile Patterns*,
https://doi.org/10.1007/978-981-96-5451-2_5

more patterns are all men's clothing, such as five patterns with NO.06, NO.31 men's sleeveless long coat; NO.13, NO.14 men's sleeveless short coat; NO.30, NO.35 men's long-sleeved long coat, six patterns with NO.36, NO.58 men's long-sleeved long coat, eight patterns with NO.05 men's sleeveless long coat. From this, the Atayal men's clothing with more pattern types is concentrated in the traditional clothing of the Klesan, Plnawan, Bunsui, and Mushas. Atayal textile patterns are mainly based on geometric patterns. Due to the many different forms of pattern shaping, a variety of woven patterns are produced. The abovementioned textile patterns can be summarized into 12 types, but each pattern has different shaping features. For example, if you take the woven pattern of the same piece of clothing, you can find a composite stripe composed of many other patterns. To analyze the shaping features of the patterns, this paper deconstructs the composite stripes of the measured textiles, removes their background color and the decoration around the patterns, and only uses a single pattern for analysis and comparison. The following continues to explain the shaping features of different patterns. The changing pattern, Fig. 5.1, shows two similar patterns in the same piece of clothing; one is a narrower shape, such as the NO.53ⓐ Nanshi; the other is slightly flat, such as the NO.53ⓒ Nanshi. Due to the combination of shapes like diamonds, circles, rectangles, etc., the pattern is formed by vertical arrangement, rich in changes and complex to define its style, so it is named after its shaping features. This pattern is relatively rare in Atayal textiles. Diamond pattern: In the 73 research samples, 63 pieces of textiles used diamonds as a decorative theme. By disassembling and analyzing their composite stripes, 220 diamond patterns can be obtained. Their performance forms can be divided into internal and external structure-shaping features. Figures 5.2, 5.3, 5.4, 5.5, 5.6, 5.7, 5.8, and 5.9 show these patterns with the internal diamond structure as the shaping feature, with single, four, nine and multiple internal diamond structures. The design elements are expressed in patterns with points, lines, crosses, diamonds, or reverse white effects. Figure 5.2 shows that the single diamond structure inside the diamond pattern is a regular diamond, among which NO.13-backⓑ Klesan, NO.26ⓒ Malepa; NO.16-frontⓒ Klesan, NO.63-backⓑ Pulibau group; NO.18ⓔ Klesan, NO.58-frontⓑ Musha, are patterns with similar structures.

As shown in Fig. 5.3, the single diamond structure within the diamond pattern is elongated, with cross-shaped elements inside the NO.05-Backⓒ Klesan, NO.68ⓑ Truku, and NO.68ⓒ Truku. Figure 5.4 shows a flat single diamond structure within the diamond pattern, including NO.10-Frontⓐ Klesan, NO.66ⓑ Truku; NO.38ⓐ1 Bunsui, NO.63-Backⓐ Pulibau, which are patterns with similar structures.

As shown in Fig. 5.5, the four-diamond structures within the diamond pattern are regular diamonds, including NO.01ⓑ Hakul (Hakku), NO.61ⓐ1 Musha; NO.14-Backⓑ Klesan, NO.51-Backⓑ Nanshi, which are patterns with similar structures. Figure 5.6 shows an elongated four-diamond structure within the diamond pattern, with cross-shaped elements inside the NO.34-Backⓑ Bunsui, NO.55-Frontⓑ2 Meicyubus. Figure 5.7 shows a flat four-diamond structure within the diamond pattern, NO.40-Frontⓐ Bunsui and NO.47-Frontⓒ Beishi, which are patterns with similar structures.

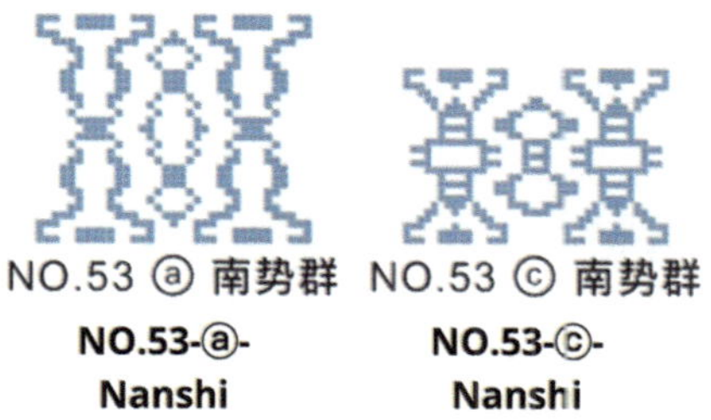

Fig. 5.1 The modeling characteristics of the changing patterns (drawn by this study) (Description: Variational patterns refer to two similar designs within the same garment, one relatively elongated and the other flatter. These patterns often combine shapes resembling diamonds, circles, or rectangles, arranged vertically to form complex patterns. Due to their diversity and difficulty categorizing them, they are named variational patterns)

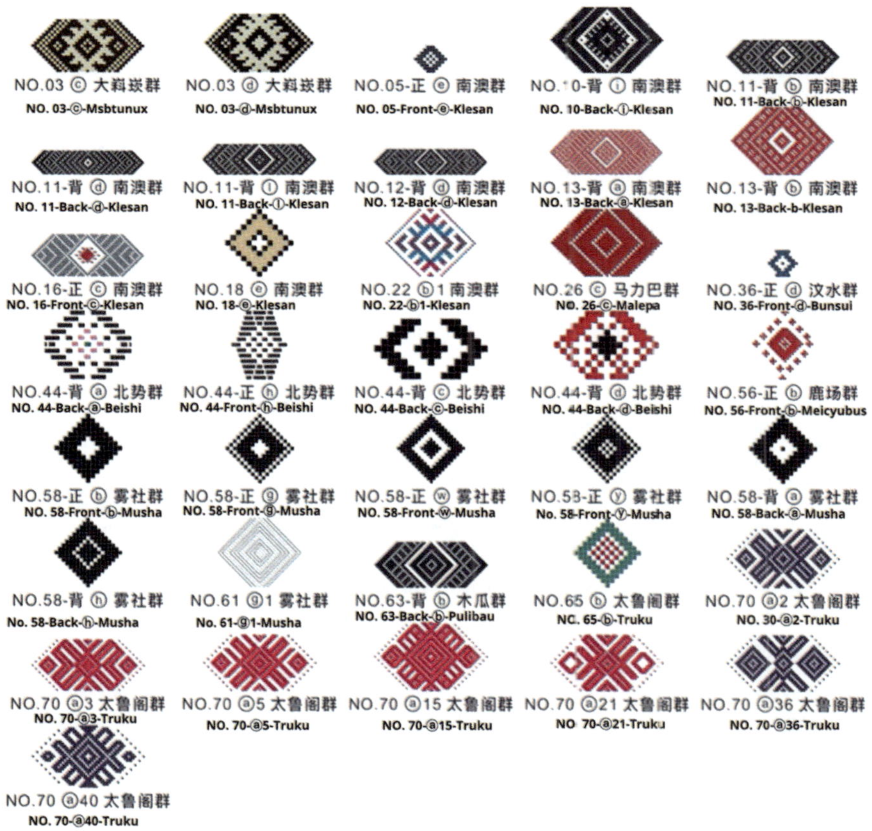

Fig. 5.2 Features of the diamond pattern (drawn by this study) (Description: The diamond pattern in this example features a single "regular diamond" structure)

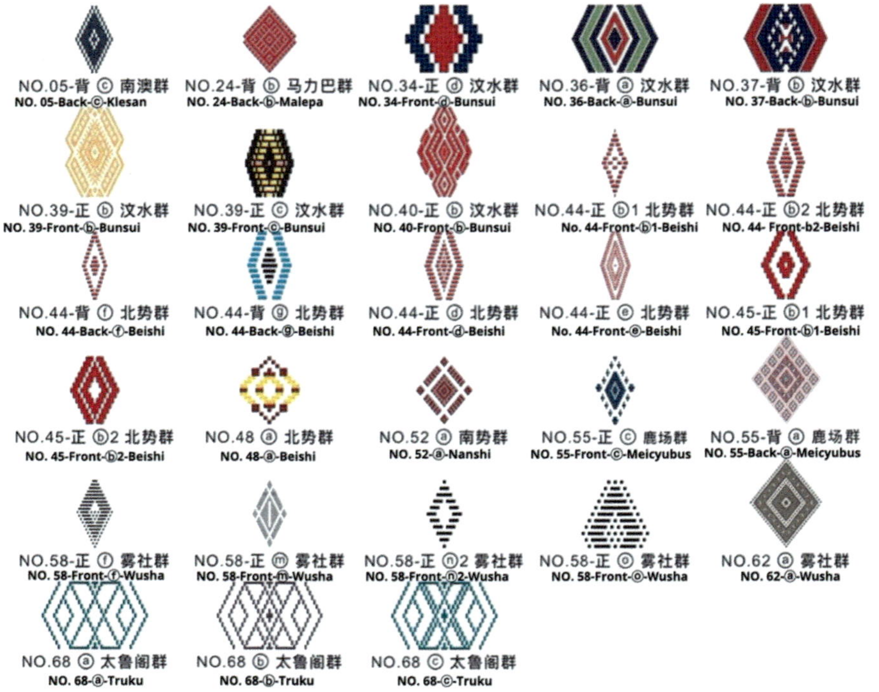

Fig. 5.3 Features of the diamond pattern (drawn by this study) (Description: The diamond pattern in this example features a single "elongated diamond" structure)

Fig. 5.4 Features of the diamond pattern (drawn by this study) (Description: The diamond pattern in this example features a single "flat diamond" structure)

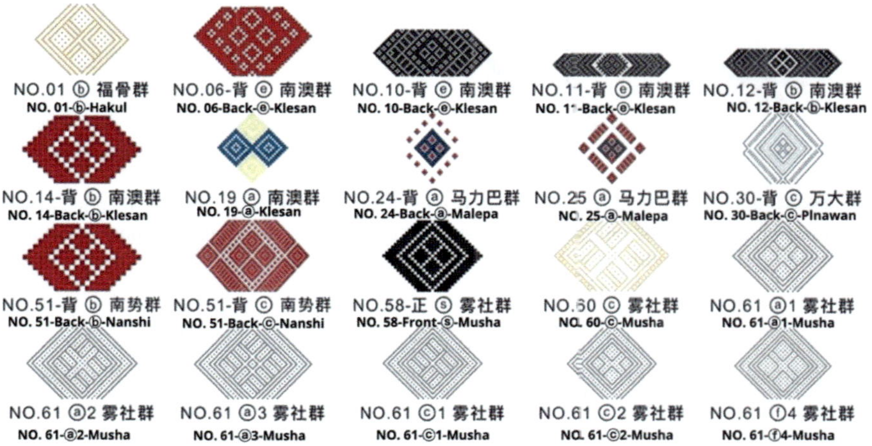

Fig. 5.5 Features of the diamond pattern (drawn by this study) (Description: The diamond pattern in this example includes four "regular diamond" structures)

Fig. 5.6 Features of the diamond pattern (drawn by this study) (Description: The diamond pattern in this example includes four "elongated diamond" structures)

Fig. 5.7 Features of the diamond pattern (drawn by this study) (Description: The diamond pattern in this example includes four "flat diamond" structures)

Fig. 5.8 Features of the diamond pattern (Description: The diamond pattern in this example features nine diamond structures, forming a dense diamond arrangement)

Fig. 5.9 Features of the diamond pattern (drawn by this study) (Description: The diamond pattern in this example consists of numerous diamond structures, with up to 30 small diamonds combined to create the textile design)

Figure 5.8 shows that the diamond pattern contains nine diamond structures; each diamond within the NO.56-Backⓑ Meicyubus comprises four small diamonds, forming many diamond organizations. Figure 5.9 shows a diamond pattern with multiple diamond structures; this structure is most representative in the pattern of NO.60ⓓ Musha, which is composed of up to 30 small diamonds combined into a woven pattern; these patterns usually appear on shawls.

Figures 5.10 and 5.11 show these patterns feature internally narrow or flat cross-shaped structures, primarily with single or four cross-shaped constructions. The design elements use points, lines, crosses, and reverse white to express the patterns. Figure 5.10 shows a single narrow, flat cross-shaped structure within the diamond pattern, where NO.06-Backⓑ Klesan and NO.44-Frontⓖ Beishi have similar pattern structures. Figure 5.11 shows four narrow, flat cross-shaped structures within the diamond pattern, with NO.26ⓑ Malepa and NO.67ⓑ Truku having similar pattern structures.

As shown in Fig. 5.12, these patterns feature internal vertical or horizontal stripe structures, with design elements using points, lines, crosses, diamonds, or reverse white to express the patterns. For example, NO.51-Frontⓑ Klesan and NO.58-Frontⓣ Musha have patterns with internal vertical stripe structures, while NO.34-Frontⓑ Bunsui, NO.42-Frontⓐ Beishi, and NO.49ⓐ Beishi have patterns with internal horizontal stripe structures.

Figures 5.13 and 5.14 show these patterns feature internal diagonal lines, or the internal diamond is diagonally divided into two, three, or four diagonal lines. Figure 5.13 shows diamond patterns featuring two internal diagonal lines, a typical pattern in Truku textiles mainly found on long skirts and shawls.

Fig. 5.10 Diamond pattern features (drawn by this study) (Description: The diamond pattern in this example incorporates a single, elongated, and flat cross-shaped structure. These patterns emphasize internal elongated or flat cross shapes, which can appear as single or four-part cross constructions. Design elements include dots, lines, crosses, and reverse effects representing patterns)

Fig. 5.11 Diamond pattern features (Description: This example features diamond patterns with four elongated and flat cross-shaped structures. The designs emphasize internal elongated or flat cross shapes, primarily in single or four-part constructions. Design elements include dots, lines, crosses, and reverse effects to create the patterns)

Fig. 5.12 Diamond pattern features (drawn by this study) (Description: The patterns in this example highlight vertical or horizontal stripe structures within the diamonds. Design elements include dots, lines, crosses, diamonds, or reverse effects, creating a balance of solid and void)

Fig. 5.13 Diamond pattern features (drawn by this study) (Description: This example showcases diamond patterns featuring internal diagonal stripes or diamonds subdivided into two, three, or four diagonal sections)

Figure 5.14 shows the internal diamond divided diagonally into two diagonal lines, such as NO.05-Backⓛ Klesan, NO.56-Backⓐ Meicyubus, NO.62ⓑ Musha, NO.64ⓐ Truku Group, with no fixed direction for the diagonal lines, expressed as either left or right. NO.05-Backⓖ Klesan, NO.12-Backⓒ Klesan, NO.12-Backⓗ Klesan, NO.61ⓓ5 Musha have the internal diamond divided diagonally into three diagonal lines. NO.62ⓒ Musha has the internal diamond divided diagonally into four diagonal lines, with a more complex pattern structure mixed with different elements such as frames, squares, etc. Usually, complex and variable patterns require a larger fabric surface to complete the weaving. This piece of textile is a mat, which is also the largest Atayal textile seen so far.

As shown in Fig. 5.15, these patterns feature an internal cross-shaped structure, using the cross in combination with a cruciform or diamond shape, presenting a

Fig. 5.14 Diamond pattern features (drawn by this study) (Description: The diamond patterns in this example are divided into two diagonal sections with no fixed orientation, displayed either leftward or rightward. Using NO.62, a textile from the Musha, as an example, the internal diamonds are divided into four diagonal sections. These complex designs often include additional elements such as frames and grids. Creating such intricate patterns typically requires a larger fabric surface, as seen in this woven mat)

symbol like "※" and displaying the pattern with a solid and hollow effect. NO.10-Backⓒ Klesan, NO.34-Frontⓐ Bunsui, NO.36-Frontⓑ Bunsui, and NO.39-Frontⓔ Bunsui, their pattern structures are similarly composed of two flat diamonds on the left and right. This feature will inevitably form when the pattern uses a cross as the core structure. Among them, NO.22ⓑ2 Klesan and NO.70ⓐ1 Truku have similar pattern structures.

As shown in Fig. 5.16, these patterns feature an internal mortar-shaped structure combined with a cruciform or diamond shape, presenting an appearance like a pestle and mortar and displaying the pattern with a solid and hollow effect. NO.22ⓑ11 Klesan, NO.41-Frontⓑ Beishi, the structure in between is composed of similar elements.

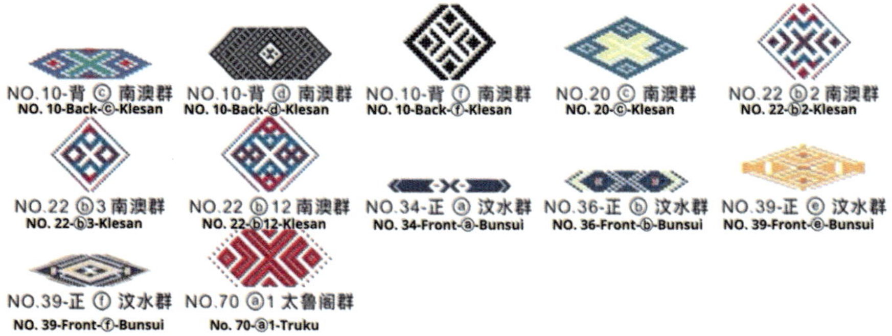

Fig. 5.15 Diamond pattern features (drawn by this study) (Description: This example features patterns with internal cross-shaped structures, combining crosses and diamonds to create symbols resembling "※." The designs utilize solid and void effects to present the patterns)

Fig. 5.16 Diamond pattern features (drawn by this study) (Description: Internal mortar-shaped structures characterize these patterns. The mortar shape is combined with crosses or diamonds, resulting in designs that resemble traditional pestle and mortar tools)

Fig. 5.17 Diamond pattern features (drawn by this study) (Description: These patterns are centered around a diamond structure, extending rectangles outward in four directions. Design elements include dots, lines, crosses, diamonds, and reverse effects to create a balanced appearance of solid and void spaces)

Figure 5.17 shows that these patterns structurally use a diamond as the core, extending rectangles in four directions. Design elements use dots, lines, cruciform, diamonds, or reverse white to display the pattern with a solid and hollow effect. NO.17ⓒ Klesan, NO.22ⓑ5 Klesan, NO.58-Backⓔ Musha, and NO.58-Backⓕ Musha have similar pattern structures. NO.44-Backⓔ Beishi extends rectangles in four directions, but the internal structure is a reverse white cross, identical to NO.22ⓑ5 Klesan.

As shown in Fig. 5.18, these patterns are composed of two diamonds arranged internally or externally, with the diamond as the shaping feature, combining its form into a pattern up and down or left and right. Design elements use dots, lines, cruciform, diamonds, or reverse white to display the pattern with a solid and hollow effect. The composition of the pattern can be summarized according to the arrangement: Separation, such as NO.11-Backⓐ Klesan. Contact, such as NO.32-Frontⓓ Bunsui, NO.72ⓐ Truku, NO.39-Backⓑ Bunsui, NO.40-Frontⓔ Bunsui. They overlap, such as NO.17ⓚ Klesan, NO.44-Frontⓐ Beishi, NO.37-Frontⓓ1 Bunsui, NO.39-Backⓓ Bunsui, and NO.45-Frontⓒ Beishi. Union, such as NO.18ⓗ Klesan, NO.44-Backⓑ Beishi, NO.08-Backⓐ Klesan, NO.41-Frontⓐ Beishi, NO.55-Frontⓑ1 Meicyubus, NO.10-Backⓑ Klesan, NO.47-Frontⓓ Beishi. Among them, NO.41-Frontⓐ Beishi has a similar octagonal flower pattern structure in the middle, which is relatively rare in Atayal textiles.

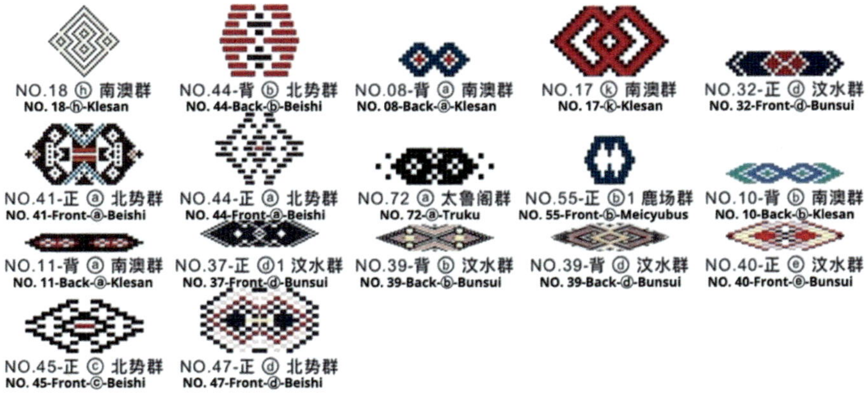

Fig. 5.18 Features of diamond patterns of group 1 (drawn by this study) (Description: These patterns consist of two diamonds arranged internally or externally. The diamonds are combined vertically or horizontally to form distinct motifs. Design elements include dots, lines, crosses, diamonds, and reverse effects to enhance the visual appeal)

As shown in Fig. 5.19, these patterns are structurally composed of vertically arranged diamonds, forming highly variable designs. Although their forms vary greatly, it is still noticeable that the patterns are centered around a diamond structure. Design elements like dots, lines, crosses, and diamonds represent the patterns. In the NO.04ⓑ3 Msbtunux, a mortar-like structure is presented in the middle. Compared to the mortar-like structure of the pattern in Fig. 5.16, this pattern is more complex.

As shown in Figs. 5.20, 5.21, and 5.22, these patterns are characterized by an external diamond structure formed by flat crosses and regular crosses combined into diamond patterns. Among them, NO.11-Frontⓐ Klesan, NO.67ⓐ Truku group, NO.44-Frontⓒ Beishi, NO.65ⓐ Truku, NO.35-Frontⓒ Bunsui, NO.72ⓑ Truku are patterns with similar structures. NO.06-Frontⓔ Klesan, NO.23ⓑ Crosses form Klesan combined into diamond patterns with a more special structural form.

Figure 5.23 shows an external diamond structure that characterizes these patterns, and the overall form appears relatively elongated. NO.33-Backⓑ Bunsui, and NO.57-Frontⓐ Musha, are patterns with similar structures; the difference in form is only the number of basic shapes.

Figure 5.24 shows that these patterns' internal and external structures are unique and difficult to categorize. Design elements like dots, lines, crosses, diamonds, or

Fig. 5.19 Features of diamond patterns of group 2 (drawn by this study) (Description: These patterns are formed by vertically arranging diamond shapes in highly variable configurations. Despite their diversity, the designs maintain a diamond-centric structural identity)

Fig. 5.20 Features of diamond patterns of group 3 (drawn by this study) (Description: These patterns feature external diamond structures created by combining flattened or regular cross shapes. The designs highlight the geometric harmony of these elements)

Fig. 5.21 Features of diamond patterns of group 4(drawn by this study) (Description: These patterns feature external diamond structures created by combining flattened or regular cross shapes. The designs highlight the geometric harmony of these elements)

Fig. 5.22 Features of diamond patterns of group 5 (drawn by this study) (Description: These patterns feature external diamond structures created by combining flattened or regular cross shapes. The designs highlight the geometric harmony of these elements)

Fig. 5.23 Features of diamond patterns of group 6 (drawn by this study) (Description: These patterns emphasize external diamond structures with an overall elongated and narrow appearance)

reverse white represent the patterns. NO.36-Frontⓒ1 Bunsui and NO.37-Frontⓐ Bunsui have similar diamond appearances, but their internal structures are very different. The internal structure of the NO.36-Frontⓒ1 Bunsui comprises two vertically arranged diamonds, while the internal structure of the NO.37-Frontⓐ Bunsui is similar to the "rice" shape. NO.44-Frontⓘ Beishi, NO.44-Frontⓙ Beishi, and NO.45-Frontⓐ Beishi, are all composed of lines forming diamond patterns, but the appearance of the form is different. The only identifiable standard part is that the inside of the pattern is all highlighted by a reverse white cross to bring out the features of the form.

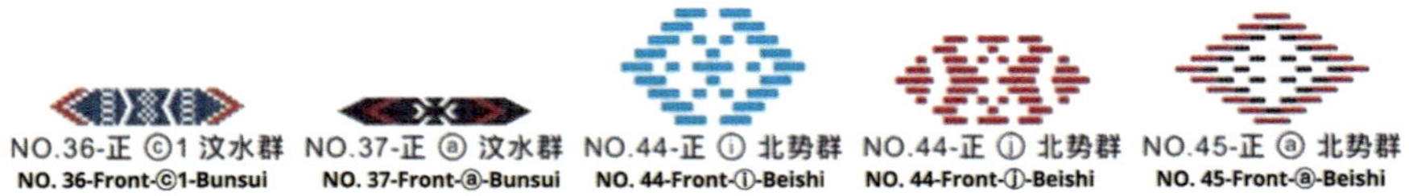

Fig. 5.24 Characteristics of diamond patterns of group 7 (drawn by this study) (Description: These patterns exhibit unique internal and external structures, making them difficult to categorize. Design elements include dots, lines, crosses, diamonds, and reverse effects to showcase a mix of solid and void spaces)

Circular patterns: among the 73 research samples, eight pieces of textiles use circles as decorative themes. Ten circular patterns can be obtained by deconstructing and analyzing their composite stripes, as shown in Fig. 5.25. Circular patterns have different forms, with diagonal lines constructing features like circular arcs. The more layers of diagonal lines, the closer they are to a circle. Among them, NO.04ⓒ Msbtunux, NO.05ⓐ Klesan, and NO.29ⓐ Taranan have similar pattern structures, their relatively simple design styles, and the woven pattern area is smaller. NO.40-Positiveⓒ Bunsui, NO.50ⓑ Beishi, and NO.50ⓓ Beishi have similar pattern structures; the overall structure is close to a perfect circle, and the circular arc line on the edge is more precise. NO.49ⓑBeishi and NO.53ⓑ Nan Shi Group have similar pattern structures but different forms. For example, the structure of NO.49ⓑ Beishi is closer to a perfect circle, while NO.53ⓑ Nan Shi Group is elongated, composed of circles arranged vertically.

Mortar-shaped patterns: among the 73 research samples, four pieces of textiles use mortar shapes as decorative themes. Four mortar-shaped patterns can be obtained by deconstructing and analyzing their composite stripes, as shown in Fig. 5.26. The characteristic of the mortar-shaped pattern is that the shape is like a rice-pounding tool. Design elements express patterns using points, lines, crosses, diamonds, or adverse space effects. The pattern structure can be divided into elongated and flat shapes, such as NO.41-Positiveⓑ Beishi, NO.45-Positiveⓓ Beishi, and NO.63-Positiveⓑ

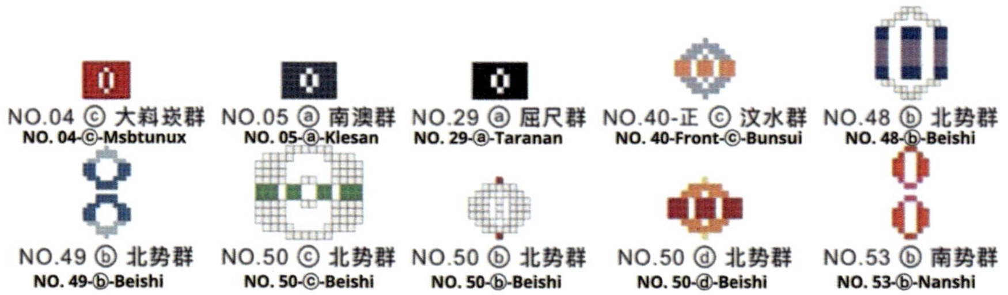

Fig. 5.25 Characteristics of circular patterns (drawn by this study) (Description: Circular patterns appear in eight textile samples from the 73 studied. Through disassembly and analysis, ten circular motifs were identified. Circular patterns vary in style, with diagonal weaves forming shapes resembling curved lines. The more diagonal layers used, the closer the design resembles a perfect circle)

Fig. 5.26 Characteristics of mortar-shaped patterns (drawn by this study) (Description: Mortar patterns use mortar shapes as decorative motifs. These designs resemble traditional rice-pounding tools)

Pulibau, all flat structures. NO.45-Positiveⓓ Beishi and NO.63-Positiveⓑ Pulibau have similar patterns. NO.65ⓑ Truku has an elongated structure, and there are no similar patterns currently.

Zigzag patterns: Of the 73 research samples, 15 textiles used zigzag shapes as decorative themes. By deconstructing and analyzing their composite stripes, 29 zigzag patterns can be obtained, as shown in Fig. 5.27. The shaping features of zigzag patterns are formed by continuous repetitive patterns in two directions. They can also be seen as primarily structured like mountains, forming zigzag patterns due to the organizational arrangement. Design elements use lines, crosses, mountains, diamonds, or negative space to express the pattern, such as NO.05-frontⓒ Klesan, NO.05-backⓔ Klesan, NO.55-frontⓒ Meicyubus, etc., presenting broader zigzag patterns in the negative space. The pattern structure can be divided into right-angled and acute-angled zigzag patterns, such as NO.03ⓐ Msbtunux, NO.10-backⓖ Klesan, NO.10-backⓗ Klesan, NO.17ⓑ Klesan, and NO.17ⓓ Klesan, all of which are right-angled zigzag patterns. NO.02-backⓐ Makanji, NO.05-frontⓒ Klesan, NO.05-backⓔ Klesan, NO.33-backⓐ Bunsui, and NO.34-frontⓒ Bunsui, all of which are acute-angled zigzag patterns. Among them, NO.17ⓓ Klesan, NO.31-backⓔ Bunsui, NO.58-backⓘ Musha; NO.18ⓖ Klesan, NO.31-backⓕ Bunsui; NO.30-frontⓐ Plnawan, NO.58-backⓓ Musha; NO.30-backⓑ Plnawan, NO.58-frontⓔ Musha, are patterns with similar structures. In addition, the rarer zigzag patterns are like NO.58-backⓚ 1 Musha and NO.58-backⓚ 2 Musha; their shaping features are straight zigzag patterns, and they present asymmetrical shapes; these two sections of weaving patterns are from the same composite stripe.

Square patterns: among the 73 research samples, six textiles use squares as decorative themes. After deconstructing and analyzing their composite stripes, 8 square patterns can be obtained, as shown in Fig. 5.28. The shaping features of square patterns are negative spaces (grids) organized within the stripes, or they can be seen as being structured by large crosses, forming square patterns due to the organizational arrangement. Design elements use dots, crosses, diamonds, or negative space to express the pattern, such as NO.05- frontⓓ Klesan, NO.13- frontⓐ Klesan, NO.16-frontⓖ Klesan, etc. The research samples found that textiles decorated with square patterns are concentrated in the Klesan, and a few similar weaving structures can also be seen in the Bunsui's textiles. The square patterns of the Klesan textiles are more significant, while the square patterns of the Bunsui textiles are smaller.

Fig. 5.27 Shaping features of zigzag patterns (drawn by this study) (Description: Zigzag patterns use zigzag shapes as decorative motifs. These patterns are analyzed and formed by repetitive two-dimensional arrangements. They can also be interpreted as mountain-shaped structures whose organizational alignment creates the zigzag effect)

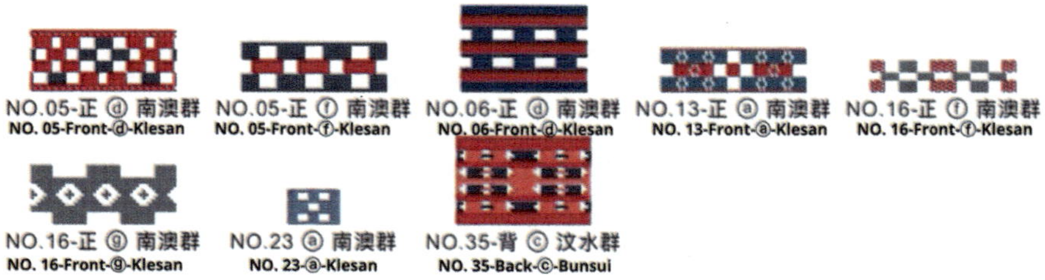

Fig. 5.28 Shaping features of square patterns (drawn by this study) (Description: Square patterns use squares as decorative motifs. Upon analysis, their defining feature is reverse squares (white squares) within the stripes, or they can be seen as large crosses arranged to form square shapes)

Mountain patterns: among the 73 research samples, 32 textiles use mountains as decorative themes. By deconstructing and analyzing their composite stripes, 49 mountain patterns can be obtained. These mountain patterns are like the expression forms of diamond patterns and can be divided into internal structure and external structure shaping features.

As shown in Fig. 5.29, these patterns are characterized by the internal mountain-shaped structure, composed of mountain and diamond shapes, such as NO.02-Back ⓐ Makanji and NO.22ⓐ Klesan. The design elements use dots, lines, crosses, diamonds, or negative space to express the patterns. The pattern structure can be divided into regular mountain shapes, elongated shapes, and flat shapes, such as NO.10-Backⓖ Klesan, NO.10-Backⓗ Klesan, NO.22ⓐ Klesan, NO.34-Frontⓑ Bunsui, and NO.58-Frontⓡ Musha, etc. are all regular mountain patterns, NO.02-Backⓐ Makanji, NO.05-Frontⓒ Klesan, NO.05-Backⓔ Klesan, NO.06-Frontⓕ Klesan, and NO.08-Backⓑ2 Klesan, etc. are elongated mountain patterns. In contrast, NO.06-Backⓒ Klesan, NO.07-Backⓑ Klesan, NO.07-Backⓓ Klesan, NO.09-Backⓐ Klesan, and NO.18ⓘ Klesan are flat mountain patterns. Among them, NO.02-Backⓐ Makanji, NO.08-Backⓑ2 Klesan, NO.34-Backⓐ Bunsui; NO.05-Frontⓒ Klesan, NO.36-Backⓑ Bunsui; NO.07-Backⓒ Klesan, NO.58-Backⓘ Musha, are respectively patterns with similar structures.

As shown in Fig. 5.30, these patterns are characterized by internal white space or other structural fillings, including rectangles, dots (squares), crosses, etc. Among them, NO.03ⓑ Msbtunux and NO.15-Frontⓐ Klesan are patterned with similar structures. Although the internal structure is different, the patterns of NO.49ⓑ Beishi and NO.56-Frontⓐ Meicyubus are identical in appearance.

As shown in Fig. 5.31, these patterns are characterized by the external mountain-shaped structure, composed of the regular mountain shape and flat shape combined

Fig. 5.29 Mountain-shaped pattern features (drawn by this study) (Description: Mountain-shaped patterns use mountain forms as decorative motifs, resembling diamond patterns in style. These patterns can be classified based on internal or external structures)

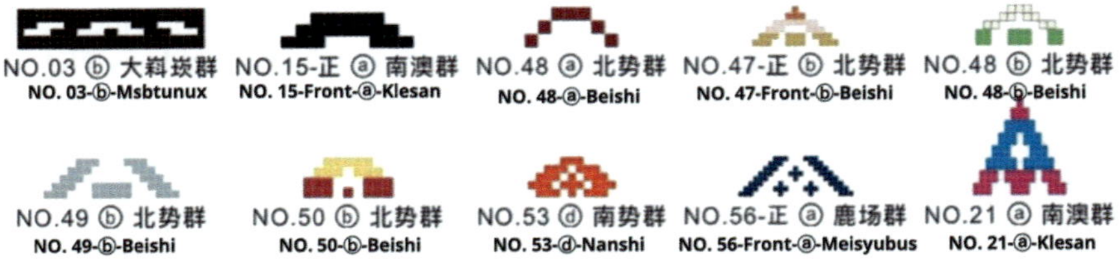

Fig. 5.30 Mountain-shaped pattern features (drawn by this study) (Description: These patterns are defined by internal blank spaces or filled structures such as rectangles, dots (squares), or crosses)

into a mountain structure. Design elements use dots, diagonal lines, or negative space to express the patterns, such as NO.08-Frontⓐ Klesan, NO.58-BackⓙMusha, NO.14-Frontⓐ Klesan, etc. NO.05-Backⓑ Klesan, NO.51-Backⓐ Nanshi; NO.08-Frontⓐ Klesan, NO.35-Frontⓐ Bunsui; NO.13-Backⓐ Klesan, NO.30-Backⓑ Plnawan; NO.26ⓐ Malipa Group, NO.31-Backⓐ Bunsui, are respectively patterns with similar structures. NO.58-Backⓙ Musha, with a mountain organization composed of diagonal lines, only appears in Musha textiles. Currently, there are no other similar patterns.

Cross patterns: Of the 73 research samples, 16 textiles used cross shapes as a decorative theme. By deconstructing and analyzing their composite stripes, 26 cross patterns can be obtained, as shown in Fig. 5.32. The cross pattern is the most common element in the weaving structure and is the basic unit of the intersection of warp and weft lines, often used in combination with large patterns or organized into smaller composite stripes. These cross patterns can be categorized based on figure-ground relationships, including solid (figure) structures, hollow (ground) structures, and other distinctive shaping features, such as NO.05-Backⓓ Klesan, NO.07-Frontⓐ Klesan, etc. Design elements display patterns using dots, crosses, or reverse white effects. The research samples found that cross-patterned textiles are concentrated in the Klesan. A few of the Plnawan, Bunsui, Nanshi, Musha, etc. also have similar weaving patterns, including NO.05-Backⓑ Klesan, NO.60ⓐ Musha;

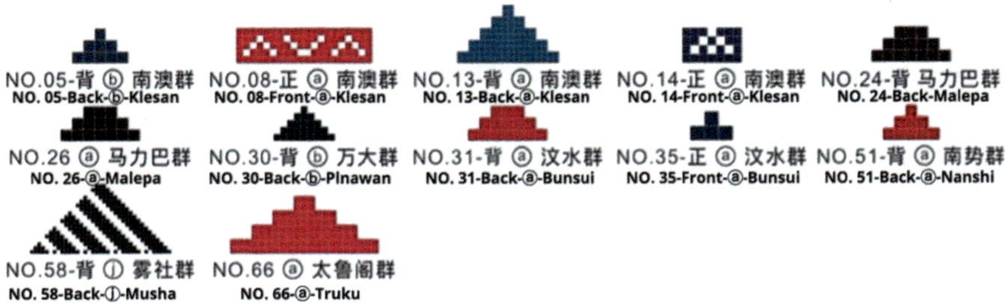

Fig. 5.31 Mountain-shaped pattern features (drawn by this study) (Description: These patterns are characterized by external mountain-shaped structures, combining regular and flat mountain forms. Design elements include dots, diagonal lines, or reverse effects to create a mix of solid and void spaces)

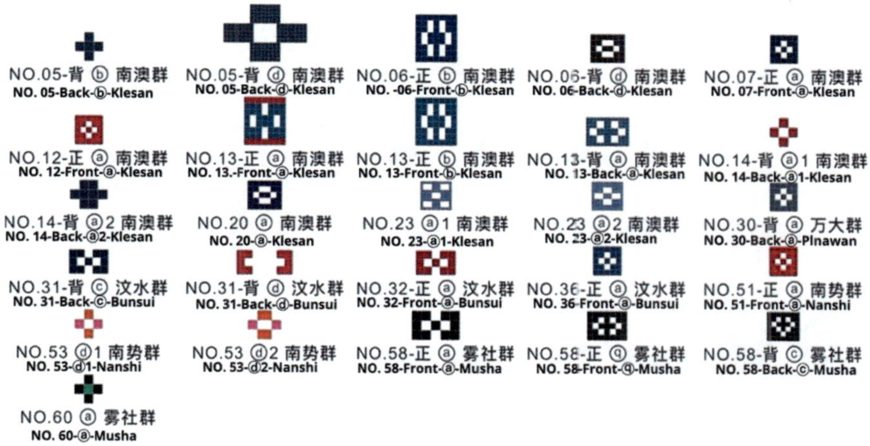

Fig. 5.32 Shaping features of cross patterns (drawn by this study) (Description: The cross pattern uses the cross shape as a decorative theme. It is also the most common element in the weaving structure and the basic unit of the intersection of warp and weft lines. It often matches large patterns or organizes them into smaller composite stripes)

NO.07-Frontⓐ Klesan, NO.30-Backⓐ Plnawan, NO.36-Frontⓐ Bunsui, NO.51-Frontⓐ Nanshi; NO.13-Backⓐ Klesan, NO.58-Frontⓖ Musha; NO.14-Backⓐ1 Klesan, NO.53ⓓ1 Nanshi; NO.31-Backⓒ Bunsui, NO.58-Frontⓐ Musha, patterns with similar structures.

X-shaped patterns: among the 73 research samples, 12 textiles use the X shape as a decorative theme. By deconstructing and analyzing their composite stripes, 18 X-shaped patterns can be obtained, as shown in Fig. 5.33. These X-shaped patterns have shaping features that can perceive the X-shaped structure within the composite stripes, or they can be regarded as two side-by-side diamond patterns, which form an X-shaped structure due to the organization and arrangement, presenting an "X◆X◆X" form. Design elements display patterns using dots, lines, crosses, diamonds, or reverse white effects. The appearance of the X-shaped pattern, such as the regular X shape, elongated shape, and flat shape, will be affected by the diamond organization on both sides, leading to changes in the appearance of the X-shaped pattern, such as NO.11-Backⓓ Klesan, NO.39-Frontⓐ Bunsui, etc. NO.07-Backⓐ Klesan, NO.11-Backⓓ Klesan, and NO.14-Frontⓒ Klesan are regular X-shaped patterns, NO.05-Backⓕ Klesan, NO.45-Frontⓓ Beishi, and NO.55-Frontⓒ Meicyubus are elongated X-shaped patterns. In contrast, NO.11-Backⓐ Klesan, NO.36-Frontⓓ Bunsui, and NO.39-Backⓒ Bunsui are flat X-shaped patterns. Among them, NO.45-Frontⓓ Beishi and NO.55-Frontⓒ Meicyubus are patterned with similar structures. Pattern structures are like NO.20ⓑ Klesan and NO.39-Frontⓐ Bunsui, which develop with diamonds as the shaping, and NO.58-Backⓡ Musha presents an asymmetrical form.

Mesh stripes: In the 73 research samples, only two textiles used mesh stripes as a decorative theme. Two mesh stripes can be obtained by dismantling and analyzing

Fig. 5.33 Shaping features of X-shaped patterns (drawn by this study) (Description: The × -shaped pattern uses × -shape as the decorative theme. The × -shaped structure can be detected within the composite stripes or as two side-by-side diamond patterns. As a result of the tissue arrangement, an × -shaped structure is formed, presenting a " × ◆ × ◆ × " shape)

their composite stripes, as shown in Fig. 5.34. The shaping feature of mesh stripes is like small grids. Design elements use points, lines, and other virtual and natural effects to express the pattern. NO.27-Rightⓐ Taranan and NO.72ⓒ Truku group have similar pattern structures.

Dot stripes: in the 73 research samples, only three textiles use dot stripes as a decorative theme. Four dot stripes can be obtained by dismantling and analyzing their composite stripes, as shown in Fig. 5.35. The shaping feature of dot stripes is that the weave is organized into stripes interspersedly. Usually, the pattern is relatively small, such as in the women's long skirt of sample NO.28 Taranan. Design elements use points, lines, and other virtual and natural effects to express the pattern. NO.29ⓐ Taranan and NO.59-Rightⓐ Musha have similar pattern structures.

Stripes: In the 73 research samples, 51 textiles used stripes as a decorative theme. The following lists 20 textiles as examples, as shown in Fig. 5.36. In addition to the composite stripes of Atayal tribal textiles composed of different patterns, the next most common is stripe organization. Its feature is to create a variety of stripe color

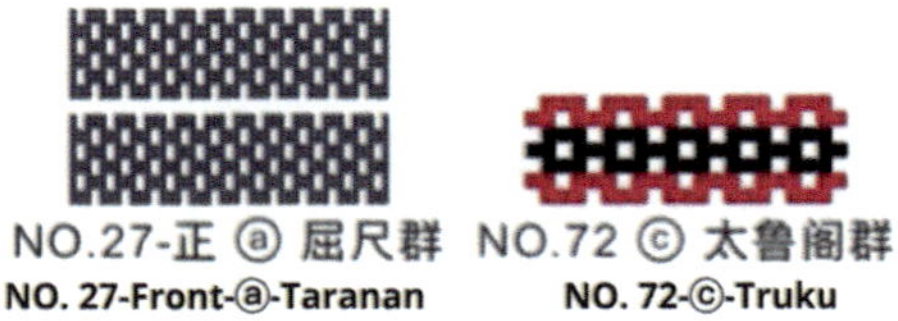

Fig. 5.34 Shaping features of mesh stripes (drawn by this study) (Description: The mesh stripes are used as decorative themes, and the shape features are like small grids. The design elements use dots, lines, and other virtual and real effects to express the pattern)

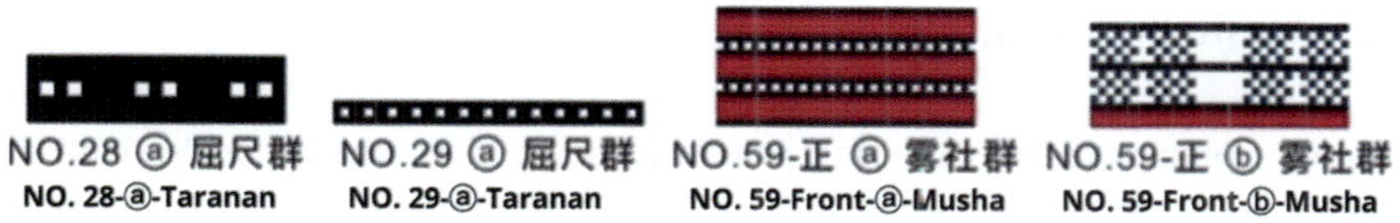

Fig. 5.35 Shaping features of dot stripes (drawn by this study) (Description: Dotted stripes are used as decorative themes. The characteristic of the shape is that the weave is organized into stripes in a grid manner, and the patterns are usually relatively small)

bands by pairing different colored threads in varying widths, such as NO.01 Hakul (Hakku), NO.03 Da Lu Kan group, NO.48 Beishi, NO.53 Nan Shi group, and NO.65 Truku, etc. Although these stripes are just combinations of colored threads, they can be summarized into similar styles of stripes from the hue or width of the stripe colors, among which NO.01 Hakul (Hakku), NO.60 Musha, NO.61 Musha, NO.40-Back-Bunsui, and NO.41-Front-Beishi are similar styles of stripes. From this, it can be known that through different colors and area arrangements of stripes, the similarity of stripes between their subgroups can also be observed, and the possible influence of their clothing forms on each other can be summarized.

Fig. 5.36 Shaping features of stripes (photographed by this study) (Description: Stripes are used as decorative themes. Composite stripes are composed of different patterns. The second most common is striped tissue, characterized by combining different colored lines with varying widths to create a variety of striped color bands)

5.2 Basic Forms of Textile Patterns

Through comparing and analyzing pattern features, if we want to understand the shape that constitutes the overall pattern further and explore how these combined elements are formed, we need to observe the relationship between parts and parts and between the whole and parts. The pattern of the research sample is studied, and the necessary basic shapes are extracted, that is, the basic unit shapes at the innermost or core of the woven pattern, for analysis and induction, as shown in Appendix 3. The constituent elements of the basic shape, whether they are main points, lines, surfaces, points and lines, lines and surfaces, regardless of the basic shape of any pattern, the parts that express the shape structure, or the elements that simplify the overall shape, all have standard features, such as repeated combinations, group combinations or arranging patterns in positive and negative spaces, or even presenting basic shapes with different hues and brightness, can be perceived through these expression techniques, and then perceive the core structure of the woven pattern.

Seven hundred seventy-four basic shapes can be extracted from the pattern of the research sample, which can be further divided into 205 types of primary forms, which can be classified into diamond, circle, rectangle, mountain, cross, × shape, diagonal shape, and other types according to the appearance of the shape. Among them, diamonds, mountains, and crosses are the basic shapes with more styles, as shown in Appendix 3 (the grey part indicates the basic shapes that appear more frequently). Count the number of times the same basic shape appears, with ten times (inclusive) or more as the quantity base. As shown in Table 5.1, it can be found that the diamond shape with 01ⓔ (11.8%) has the most basic shapes, distributed in the Makanji, Msbtunux, Klesan, Malepa, Bunsui, Beishi, Nanshi, Musha, Pulibau, and Truku. 07ⓒ (2.7%) diamond basic shape is the second, distributed in the Msbtunux, Klesan, Musha, and Truku. The following diamond basic shapes are 01ⓓ (2.6%) distributed in the Klesan, Musha, and Truku, and 20ⓒ (2.6%) distributed in the Hakul (Hakku), Musha. In addition, 19ⓐ (1.9%) and 19ⓑ (2.2%) diamond basic shapes are not few. These two basic shapes are similar, but there is just a difference in the left and right shapes. This basic shape is currently only visible in the men's and women's shawls (NO.60, NO.61) of the Musha. The basic form 31ⓓ (1.3%) is the most common in the mountain basic form, distributed in the Klesan, Bunsui, Nanshi, Musha, and Pulibaus. The cross shape with 41ⓐ (11.5%) has the most basic shapes, distributed in the Makanji, Msbtunux, Klesan, Beishi, Nanshi, Meicyubus, Musha, and Truku. 45ⓐ (9%) cross basic shape is the second, distributed in the Msbtunux, Klesan, Plnawan, Bunsui, Nanshi, Musha, Pulibao, and Truku. The next cross basic shape is 42ⓔ (2.2%) distributed in the Makanji, Klesan, Bunsui, Meicyubus, and Truku.

Although the decorative patterns of different subgroups' costumes vary, the primary forms of pattern shaping show a systematic phenomenon of similarity, such as rhombus 01ⓔ; cross shape 41ⓐ, 45ⓐ, and other primary forms, which can generally be seen in the traditional costumes of the Atayal tribe such as the Mabtunux, Klesan, Nanshi, Musha, Truku, etc., with the same primary forms. Furthermore,

Table 5.1 Distribution of primary forms in the shaping of Atayal tribe textile pattern

Graphic Type	Rhombus									Yamagata	Cross			
NO.	01 ⓓ	01 ⓔ	05 ⓑ	07 ⓒ	13 ⓔ	19 ⓐ	19 ⓑ	20 ⓒ	20 ⓔ	31 ⓓ	41 ⓐ	42 ⓔ	43 ⓑ	45 ⓐ
Basic form / Subgroup	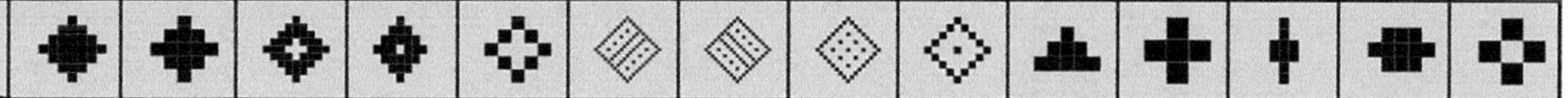													
Hakul (Hakku)					1			2	1					
Makanji		1									1	1		
Msbtunux		15		7	1						14			7
Klesan	5	27	10	5	11				4	3	31	10	1	28
Malepa		3												
Taranan														
Plnawan														2
Bunsui		6								2		1	3	5
Beishi		3									6		3	
Nanshi		2								1	3		1	1
Meicyubus											2	3		
Musha	5	5		1		15	17	18	7	3	4		3	7
Pulibau		4								1			3	2
Truku Group	10	25		8							28	2	1	18
Occurrence	20	91	10	21	13	15	17	20	12	10	89	17	15	70
Ratio (%)	2.6	11.8	1.3	2.7	1.7	1.9	2.2	2.6	1.6	1.3	11.5	2.2	1.9	9.0

Description: 774 basic shapes were extracted from the patterns of 73 Atayal textiles, which can be further divided into 205 basic forms. According to the appearance of the shapes, they can be summarized into a rhombus, circle, rectangle, mountain, cross, ×, oblique line, and other types. The most common basic shapes are rhombus, mountain, and cross. Table 5.1 shows the basic shapes of Atayal textile patterns, which can be regarded as the core elements of the patterns. In addition to their similarity, these elements are small modules in the pattern shape (Drawn for this study) some textile patterns seem complex and varied but comprise a few simple primary forms. For example, the basic form of the pattern of men's and women's shawls NO.01ⓐⓑ of the Hakul (Hakku) is 13ⓔ, 20ⓒ, 20ⓔ. The basic form of the pattern of men's and women's shawls NO.04ⓑ 1–13 of the Mabtunux is 01ⓔ, 07ⓒ, 41ⓐ, 45ⓐ. The basic form of the pattern of men's sleeveless long gowns NO.11-Backⓑⓒⓓⓔⓕⓖⓗⓘⓙⓚⓛ of the Klesan is 01ⓓ, 01ⓔ, 41ⓐ, 45ⓐ. The basic form of the pattern of men's sleeveless long gowns NO.32-Frontⓐⓑⓒⓓ, NO.32-Backⓐⓑ of the Bunsui is 01ⓔ, 43ⓑ, 45ⓐ. The basic form of the pattern of woven mats NO.62ⓐⓑⓒ of the Musha is 03ⓓ, 03ⓕ, 07ⓑ, 12ⓐ, 12ⓒ, 12ⓓ, 42ⓓ, 43ⓑ, 46ⓐ, and 55ⓒ. The shaping of the Atayal tribe's textile patterns, although composed of different combinations to form a myriad of costume patterns, is the basic form that can be said to be the core element of the patterns. These elements not only have similarities but are also small modules in pattern shaping. Overall, the basic form represents the style foundation of the textile pattern.

5.3 Textile Composite Stripe Structure

The organizational structure of textile patterns, that is, the layout of the pattern, is also the inherent form of the overall pattern. The composite stripes composed of Atayal tribal textile patterns can be classified into single (single shape), vertical separation (vertically arranged identical shaping elements with distance), horizontal separation (horizontally arranged identical shaping elements with distance), vertical repetition (vertically arranged identical shaping elements without distance), horizontal repetition (horizontally arranged identical shaping elements without distance), vertical intersection (overlapping identical shaping elements vertically), horizontal intersection (overlapping identical shaping elements horizontally), vertical connection (vertically welding identical shaping elements), horizontal connection (horizontally welding identical shaping elements), grouping (shaping components with common characteristic elements), and X-shaped (arranged in a continuous square based on the X-shape), as shown in Table 5.2. By organizing the research samples of composite stripes, 362 groups can be obtained. Still, the same group of composite stripes may present two different organizational structures, such as the composite stripes of the NO.29ⓐ group, or show similar organizational structures, in which case they can be merged into the same composite stripe, such as the composite stripes of the NO.68ⓐⓑⓒ, NO.69ⓐⓑ, the NO.71ⓐⓑⓒ groups, so there are a total of 363 groups of composite stripes available for structural analysis.

The horizontally separated composite stripes are the most numerous organizational structures in the research samples, with 240 groups accounting for 66% of the total, as shown in Table 5.3. These composite stripes are distributed in the textiles of all subgroups in the research samples, such as the NO.01ⓐ Hakul (Hakku) (men's and women's shawls), the NO.02-Frontⓐ Makanji (men's sleeveless long clothes), NO.03ⓑⓒⓓ Msbtunux (men's and women's shawls), the NO.15-Frontⓐ Klesan (men's and women's long-sleeved short clothes), the NO.26ⓐⓑⓒⓓ Malepa

Table 5.2 Methods of structural analysis of Atayal tribal textile composite stripes

Type Name	Explanation	Legend	Sample-Pattern number/group	Composite stripes	Basic shape
Single	Single shape		NO.17ⓐKlesan		
Vertical separation	The same formative elements are arranged vertically with a distance between them.		NO.39-BackⓐBunsui		
Horizontal separation	The same formative elements are arranged horizontally with some distance between them.		NO.03ⓒMsbtunux		
Vertical repetition	The same formative elements are arranged vertically without any distance.		NO.33-BackⓑBunsui		
Horizontal repetition	The same formative elements are arranged horizontally without any distance.		NO.01ⓑHakul (Hakku)		
Vertical intersection	The same formative elements overlap vertically.		NO.62ⓐMusha		
Horizontal intersection	The same formative elements overlap horizontally.		NO.18ⓗKlesan		
Vertical connection	The same shaping elements are welded longitudinally.		NO.56-BackⓐMeicyubus		
Horizontal connection	The same shape elements are welded horizontally.		NO.39-FrontⓔBunsui		
Grouping	The components of the shape have common characteristic elements.		NO.24-BackⓐMalepa		
×shape	The four sides are arranged continuously with X shape as the basic structure.		NO.52ⓐNanshi		

(tobacco bags), the NO.28ⓐ Taranan (women's long skirts), the NO.30-Frontⓐ Plnawan (men's long-sleeved long garments), the NO.31-Backⓐⓑⓒⓓ Bunsui (men's sleeveless long clothes), the NO.44-Frontⓒⓓⓔⓕⓖⓗⓘⓙ Beishi (women's long-sleeved long clothes), the NO.51-Frontⓑⓒ Nanshi (men's long-sleeved long clothes), the NO.56-Frontⓐ Meicyubus (men's sleeveless long garments), the NO.58-Frontⓐⓑⓕⓗⓘⓙⓚ ⓛⓝⓞⓟⓢⓣⓤⓦ Musha (men's long-sleeved long clothes), the NO.63- Frontⓑⓒⓓⓔ Pulibau (men's sleeveless long clothes), the NO.65ⓑ Truku (women's sleeve sets) etc. all have this structure. The horizontally separated structure is arranged horizontally with the same formative elements at a distance, using deformation (irregular form), diamond shape, mortar shape, square, mountain shape, cross, cross shape, diagonal stripe, stripe or leaving empty positions to separate the main pattern.

The compound stripes of the transverse repetitive structure are second, with 31 groups accounting for 8.5% of the total, as shown in Table 5.3. They are distributed in the textiles of subgroups such as NO.01ⓑ Hakul (Hakku) (men's and women's shawls), NO.03ⓐ Msbtunux (men's and women's shawls), NO.10-Backⓐⓗⓘ Klesan (men's sleeveless long clothes), NO.30-Frontⓑ Plnawan (men's long-sleeved

Table 5.3 Distribution of composite stripe structures in Atayal tribal textile

Type Name	Composite stripe structure											Sum
	Single	Vertical separation	Horizontal separation	Vertical repetition	Horizontal repetition	Vertical intersection	Horizontal intersection	Vertical connection	Horizontal connection	Grouping	× shape	
Subgroup												
Hakul (Hakku)			1		1							2
Makanji			3									3
Msbtunux			4		3							7
Klesan	10	3	93		8		6		2			122
Malepa			4							2	1	7
Taranan			2							3		5
Plnawan			4		1							5
Bunsui		3	39	1	10		1		3	3		60
Beishi			27		2				8	4		41
Nanshi			9		1				1		1	12
Meicyubus	1		1					2	3	1		8
Musha		1	39	2	2	2	2		4	7	4	63
Pulibau			8		2				3			13
Truku			6		1				2	5	1	15
Occurrence	11	7	240	3	31	2	9	2	26	25	7	363
Ratio (%)	3.0	1.9	66.0	0.8	8.5	0.6	2.5	0.6	7.2	6.9	1.9	100

Description: Horizontally separated composite stripes are the most common tissue structure in the study samples, with 240 groups accounting for 66% of the total (Drawn by this study)

long clothes), NO.31-Backⓔⓕ Bunsui (men's sleeveless long clothes), NO.44-Backⓓ Beishi (women's long-sleeved long clothes), NO.54ⓐ Nanshi (women's foot protection cloth), NO.58-Frontⓖⓨ Musha (men's long-sleeved long clothes), NO.63-Frontⓐ Pulibau (men's sleeveless long clothes), NO.67ⓑ Truku (women's set sleeves) and so on. This structural feature is characterized by the horizontal arrangement of the same formative elements without distance, with diamond patterns, zigzag patterns, square patterns, mountain patterns, × patterns, and other forms as the main patterns. Among them, the diamond pattern, zigzag pattern, mountain pattern, × pattern's organizational structure, due to the inevitable phenomenon of structural combination, produces diamond or mountain-shaped interval forms, and the relationship between the primary and secondary patterns can be identified by observing the size or color difference.

Next, the transverse connection structure has 26 groups accounting for 7.2% of the total, as shown in Table 5.3. They are distributed in the textiles of subgroups such as NO.18ⓔ Klesan (women's belt), NO.34-Frontⓑ Bunsui (men's and women's long-sleeved long clothes), NO.41-Frontⓐ Beishi (men's long-sleeved long clothes), NO.51-Frontⓐ Nanshi (men's long-sleeved long clothes), NO.55-Frontⓑⓒ Meicyubus (men's sleeveless long clothes), NO.58-Frontⓔⓡⓥ Musha (men's long-sleeved long clothes), NO.63-Backⓐⓑⓔ Pulibau (men's sleeveless long clothes), NO.65ⓐ Truku (women's set sleeves) and so on. This structural feature is characterized by the horizontal welding of the same formative elements, with diamond patterns, circular patterns, mountain patterns, cross patterns, net-like stripes, and other forms as the main patterns. These compound stripes are most complex in the Beishi textiles, and the structure of the woven patterns is varied, such as NO.41-Frontⓐ, NO.42-Frontⓐ (men's long-sleeved long clothes); NO.44-Frontⓔ, NO.44-Backⓔ, NO.45-Frontⓒ (women's long-sleeved long clothes); NO.49ⓐ (women's foot protection cloth).

The compound stripes of the group structure have another 25 groups accounting for 6.9% of the total, as shown in Table 5.3. They are distributed in the textiles of subgroups such as NO.24-Backⓐⓑ Malepa (men's sleeveless long clothes), NO.27-Frontⓐⓑ Taranan (men's sleeveless long clothes), NO.36-Frontⓐ Bunsui (men's long-sleeved long clothes), NO.44-Frontⓑ Beishi (women's long-sleeved long clothes), NO.55-Frontⓐ Meicyubus (men's sleeveless long clothes), NO.58-Frontⓓⓠⓧ Musha (men's long-sleeved long clothes), NO.68ⓐⓑⓒ Truku (women's long skirt) and so on. This structural feature is that the formative components have common feature elements, with diamond patterns, circular patterns, cross patterns, net-like stripes, and other forms as the main patterns, and are not limited to the area of the woven pattern, even the most miniature compound stripe composed of cross patterns also has the performance of group structure. Among them, the Truku textiles are particularly distinctive, such as NO.68, NO.69 (women's long skirt), NO.70, NO.71, and NO.73 (men's and women's shawls). Although the pattern structure is composed in a group way, there are some slight differences in the pattern formation of the same piece of textile, thus producing a collection of similar form elements.

The cross-shaped structure of the compound stripes, which only accounts for 1.9% of the total, is shown in Table 5.3. It is distributed in the textiles of subgroups such as NO.25ⓐ Malepa (men's chest cover), NO.52ⓐ Nanshi (men's chest cover), NO.58-ⓒⓜ Musha (men's long-sleeved dress), NO.64ⓐ Truku (men's chest cover), etc. This structural feature is based on the cross-shaped structure arranged in four continuous directions, with diamond and cross patterns as the main patterns. The overall structure seems to be grouped, but the arrangement is highly regular, all following the cross-shaped framework to extend the same pattern in all directions.

Single (3%), horizontal intersection (2.5%), vertical separation (1.9%), vertical repetition (0.8%), vertical intersection (0.6%), vertical connection (0.6%), all of these are less common compound stripe structures, as shown in Table 5.3. The single structure of the compound stripes is distributed in the textiles of NO.17ⓐⓒⓔⓕⓖⓗⓘⓙⓚⓛ Klesan (women's belt) and NO.55-ⓐ Meicyubus (men's sleeveless dress). This structural feature is represented by a single pattern, with diamond patterns as the main pattern, and the woven pattern area is also more prominent, especially the pattern of NO.55-ⓐ Meicyubus, which almost covers the entire back of the garment. The horizontal intersection structure of the compound stripes is distributed in the textiles of NO.10-ⓓ Klesan (men's sleeveless dress), NO.11-ⓖⓛ Klesan (men's sleeveless dress), NO.18ⓕⓗⓚ Klesan (women's belt), NO.39-ⓑ Bunsui (women's long-sleeved dress), NO.58-ⓔⓕ Musha (men's long-sleeved dress). This structural feature overlaps horizontally with the same pattern elements, with diamond patterns as the main pattern, and another diamond is produced at the horizontal overlap. The vertical separation structure of the compound stripes is distributed in the textiles of NO.17ⓑⓓ Klesan (women's belt), NO.18ⓖ Klesan (women's belt), NO.33-ⓐ Bunsui (men's sleeveless dress), NO.39-ⓐ Bunsui (women's long-sleeved dress), NO.40-ⓑ Bunsui (women's long-sleeved dress), NO.62ⓑ Musha (woven mat). This structural feature is arranged vertically with the same pattern elements at a distance, with diamond and zigzag patterns as the main patterns. Among them, the compound stripe structure of NO.39-ⓐ Bunsui and NO.62ⓑ Musha is more complex, showing a rich sense of hierarchy in form. The vertical repetition structure of the compound stripes is distributed in the textiles of NO.33-ⓑ Bunsui (men's sleeveless dress) and NO.57-ⓐⓑ Musha (men's sleeveless dress). This structural feature is arranged vertically with the same pattern elements without distance. Diamond patterns are the main pattern, and the entire textile pattern is placed vertically. The vertical intersection structure of the compound stripes is only found in the textiles of NO.62ⓐⓒ Musha (woven mat). This structural feature overlaps vertically with the same pattern elements, with diamond patterns as the main pattern, and the entire textile's pattern area is more significant. The vertical connection structure of the compound stripes is also only found in the textiles of NO.56-ⓐⓑ Meicyubus (men's sleeveless dress). This structural feature is welded vertically with the same pattern elements, with diamond patterns as the main pattern, the entire textile's pattern area is more prominent, and the pattern structure is highly complex.

The complex patterns of the Atayal tribe's textiles reveal considerable diversity and complexity, yet these patterns are composed of many similar primary forms. As shown in Table 5.2, NO.17ⓐ has a single structure, the basic form is Appendix 3's

13ⓔ, composed of 1 basic form; NO.39-Backⓐ has a vertical separation structure, the basic form is Appendix 3's 04ⓐ, 10ⓓ, composed of 2 primary forms; NO.03ⓒ has a horizontal separation structure, the basic form is Appendix 3's 13ⓔ, 45ⓐ, composed of 2 primary forms; NO.33-Backⓑ has a vertical repetition structure, the basic form is Appendix 3's 03ⓖ, composed of 1 basic form; NO.01ⓑ has a horizontal repetition structure, the basic form is Appendix 3's 20ⓒⓔ, composed of 2 primary forms; NO.62ⓐ has a vertical intersection structure, the basic form is Appendix 3's 12ⓐ, 07ⓑ, 42ⓓ, 46ⓐ, 43ⓑ, composed of 5 primary forms; NO.18ⓗ has a horizontal intersection structure, the basic form is Appendix 3's 13ⓔ, composed of 1 basic form; NO.56-Backⓐ has a vertical connection structure, the basic form is Appendix 3's 03ⓕ, 42ⓓ, composed of 2 primary forms; NO.39-Frontⓔ has a horizontal connection structure, the basic form is Appendix 3's 43ⓔ, composed of 1 basic form; NO.24-Backⓐ has a group structure, the basic form is Appendix 3's 01ⓔ, composed of 1 basic form; NO.25ⓐ has a cross structure, the basic form is Appendix 3's 07ⓑ, 03ⓕ, composed of 2 primary forms.

From these complex patterns, we can see the universality of the primary forms. For example, the basic form of the NO.17ⓐ single structure pattern is the same as that of the NO.03ⓒ horizontal separation structure and the NO.18ⓗ horizontal intersection structure. Similarly, the primary form of the NO.56-Backⓐ vertical connection structure pattern is the same as that of the NO.52ⓐ cross structure. The relationship between the complex stripe structure and pattern formation of the listed textiles lies in the fact that the pattern's basic form is the core. As the weaving process expands, it forms a pattern with a high degree of variation. Patterns composed of one or more basic forms will form more layered pattern changes or complex structures. The original basic form, influenced by the arrangement of the structure, such as direction, angle, position, and other progressive changes, will gradually disappear and merge into the pattern. Still, it may also be highlighted by color matching. Therefore, the arrangement of the pattern structure depends on the gradual changes of the basic form expanding outward.

5.4 Textile Complex Stripe Colors

Cultural background factors lead to lifestyle-determining color expression methods, especially in the natural and cultural environment formed over a long period, independently forming different ethnic groups' color forms and concepts. The preference for color use by Taiwan's minority ethnic groups can generally be seen in their material culture and clothing. Speaking of the color performance of the Atayal tribe's textiles, and according to previous related literature, it is pointed out that the rich colors, like a rainbow, symbolize the meaning of the Atayal tribe's clothing connecting to the Rainbow Bridge (Spirit Bridge) legend. The multi-colored rainbow-like clothing colors are not all characteristics of the traditional Atayal tribe's clothing. This description is closer to the clothing style of the Bunsui or Beishi, especially the bride's long-sleeved dress.

A color combination analysis was conducted on 73 pieces of Atayal tribe textiles, revealing that colorless black, grey, and white, as well as red, peach, pink, orange, yellow, green, blue, purple, brown, etc., are all commonly used in Atayal tribe textiles, as shown in Appendix 4. Here, an overall analysis and summary of the main color series of Atayal tribe textiles is first conducted, detailed as follows:

(1) The black series is likely formed by burying the thread in deeper soil colors for some time or by using the bark of the chinaberry tree as a dye, such as NO.28 Taranan (women's long skirts), NO.29 Taranan (men's and women's shawls).
(2) The grey series is close to the original color of ramie material or formed by different degrees of soaking and dyeing, resulting in varying lTable displaying different pattern types with explanations, legends, sample-pattern numbers or groups, composite stripes, and basic shapes. Types include Single, Vertical and Horizontal Separation, Repetition, Intersection, Connection, Grouping, and X-shape. Each type is associated with a unique legend symbol, sample-pattern identifier, and visual representation of composite stripes and basic shapes. Patterns are visually distinct, showcasing various arrangements and overlaps of shapes.evels of grey, such as the NO.67 Truku (women's sleeved tops). In addition, it can also be seen paired with other color series, such as NO.02 Makanji (men's sleeveless long clothes) main color series are grey, peach red; NO.13 Klesan (men's sleeveless short clothes) main color series are grey, red; NO.48 Beishi (women's foot protection cloth) main color series are grey, blue, etc.
(3) The white series is made from bleached ramie, with the Truku textiles being the most representative, such as NO.65 NO.66 women's sleeved tops; NO.68 women's long skirts, and some use pearl material strung into stripe organization, such as NO.54 Nanshi (women's foot protection cloth).
(4) The red series textiles include NO.03 Msbtunux (men's and women's shawls), NO.30 Wan Da group (men's long-sleeved long clothes), NO.33 Bunsui (men's sleeveless long clothes), NO.40 Bunsui (women's long-sleeved long clothes), NO.44 Beishi (women's long-sleeved long clothes), NO.52 Nanshi (men's chest cover), NO.57 Musha (men's sleeveless long clothes), NO.64 Truku (men's chest cover).
(5) The peach red series is paired with other color series, such as NO.02 Makanji (men's sleeveless long clothes) main color series are grey, peach red; NO.10 Klesan (men's sleeveless long clothes) main color series are black, peach red; NO.21 Klesan (men's and women's shawls) main color series are black, peach red, yellow; NO.22 Klesan (men's and women's shawls) main color series are black, peach red.
(6) The pink series is used to embellish patterns, such as NO.35, NO.36 Bunsui (men's long-sleeved long clothes), NO.39, NO.40 Bunsui (women's long-sleeved long clothes), NO.42 Beishi (men's long-sleeved long clothes), NO.43 Beishi (men's open-back trousers), NO.44 Beishi (women's long-sleeved long clothes), NO.46 Beishi (women's long-sleeved short garments).

(7) The orange series is paired with other color series, such as the main color series of the NO.38 leading Bunsui group (men's chest cover) is black and orange; the primary color series is the NO.39 Bunsui (women's long-sleeved long clothes) main color series is red.

(8) The yellow series is paired with other color series, such as the NO.19 Klesan (women's belts) main color series is yellow blue; the NO.20 Klesan (women's foot protection cloth) main color series is yellow, blue; the NO.21 Klesan (men's and women's shawls) the main color series are black, peach red, and yellow; the NO.49 Beishi (women's foot protection cloth) main color series are black, grey, and yellow; the NO.50 Beishi (men's and women's shawls) main color series are yellow and blue.

(9) The green series is paired with other color series, with only the NO.70 Truku group (men's and women's shawls) main color series being white, red, and green.

(10) The blue series is paired with other color series, such as the NO.07 Klesan (men's sleeveless long clothes) main color series being grey, red, and blue; the NO.16 Klesan (men's long-sleeved short clothes) main color series being red and blue; the NO.17 Klesan (women's belts) main color series being black, red, and blue; the NO.19 Klesan (women's belts) main color series being yellow and blue; the NO.20 Klesan (women's foot protectors) main color series being yellow and blue; the NO.23 Klesan (men's and women's shawls) main color series being red and blue; NO.24 Malepa (men's sleeveless long clothes) main color series being grey, red, and blue; the NO.25 Malepa (men's chest covers) main color series being red, and blue; the NO.27 Taranan (men's sleeveless long clothes) main color series being grey and blue; the NO.36 Bunsui (men's long-sleeved long clothes) main color series being black, grey, and blue; the NO.48 Beishi (women's foot protectors) main color series being grey and blue; the NO.50 Beishi (men's and women's shawls) main color series being yellow and blue; the NO.71 and the NO.73 Truku (men's and women's shawls) main color series being white, red, and blue.

(11) The purple series is used for embellishing patterns, such as the NO.22 Klesan (men's and women's shawls), the NO.40 Bunsui (women's long-sleeved long clothes), the NO.47 Beishi (women's long-sleeved short garments), the NO.48 Beishi (women's foot protectors), the NO.68 Truku (women's long skirts), the NO.71 Truku (men's and women's shawls).

(12) The brown series is likely dyed with sweet potato tubers, but it may also incorporate other materials to deepen the color, such as ash juice. The brown main color series only includes the NO.62 Musha (woven mats), which is paired with other color series, such as the NO.18 Klesan (women's belts), the main color series being black and brown.

The above list of color series used in Atayal tribe textiles can be summarized as black, grey, white, red, peach red, orange, yellow, green, blue, brown, etc. However, the color combinations between different main color series can still be further explored. As described below:

(1) Pairings of the primary color are black.

Black, grey, and red: the NO.12 Klesan (men's sleeveless long clothes), the NO.26 Malepa (tobacco bag), the NO.34 Bunsui (men's and women's long-sleeved long clothes), the NO.37 Bunsui (men's long-sleeved short garments), the NO.41 Beishi (men's long-sleeved short clothes), the NO.42 Beishi (men's long-sleeved short garments), the NO.43 Beishi (men's open-back trousers), the NO.46 Beishi (women's long-sleeved short garments), the NO.63 Pulibau (men's sleeveless long clothes), the NO.72 Truku (men's and women's shawls).

The NO.49 Beishi (women's foot protection cloth) is black, grey, and yellow. The NO.36 Bunsui (men's long-sleeved clothes) is black, grey, and blue. Black, white, and red: the NO.35 Bunsui (men's long-sleeved long garments) and the NO.47 Beishi (women's long-sleeved long clothes).

Black and red: the NO.01 Hakul (Hakku) (men's and women's shawls), the NO.04 Msbtunux (men's and women's shawls), the NO.11 Klesan (men's sleeveless long clothes), the NO.15 Klesan (men's and women's long-sleeved short clothes), the NO.45 Beishi (women's long-sleeved long garments), the NO.53 Nanshi (women's long skirt), the NO.58 Musha (men's long-sleeved long clothes), the NO.59 Musha (women's long-sleeved short garments); the NO.60 and the NO.61 Musha (men's and women's shawls).

Black, red, and blue: the NO.17 Klesan (women's belt).

Black and peach red: the NO.10 Klesan (men's sleeveless long clothes), the NO.22 Klesan (men's and women's shawls).

Black, peach red, and yellow: the NO.21 Klesan (men's and women's shawls).

Black, orange: the NO.38 Bunsui (men's chest cover).

Black, brown: the NO.18 Klesan (women's belt).

Among them, the textiles with the main color series of black, grey, and red are concentrated in the Klesan, Malepa, Bunsui, Beishi, Pulibau, and Truku, with the grey color series as the base cloth color. The textiles with the main color series of black and red are concentrated in the Hakul (Hakku), Msbtunux, Klesan, Beishi, Nanshi, and Musha, as the entire piece of textile is full of patterns, thus covering the base cloth color. The textiles of these two main color series combinations are in larger quantities.

(2) Pairings of the main color grey.

Grey and red: the NO.05, NO.06, NO.09 Klesan (men's sleeveless long clothes); the NO.13 Klesan (men's sleeveless short garments); the NO.31, NO.32 Bunsui (men's sleeveless long garments); the NO.51 Nanshi (men's long-sleeved long clothes); the NO.55 and the NO.56 Meicyubus (men's sleeveless long garments).

Grey, red, and blue: the NO.07 Klesan (men's sleeveless long clothes) and the NO.24 Malepa (men's sleeveless long garments).

Grey and peach red: the NO.02 Makanji (men's sleeveless long garments).

Grey and blue: the NO.27 Taranan (men's sleeveless long clothes), the NO.48 Beishi (women's foot protection cloth).

Excluding textiles with black as the main color series, it is found that textiles with grey and red as the main color series are more numerous, concentrated in the Klesan, Bunsui, Nanshi, and Meicyubus.

(3) Pairings of the white primary color scheme.

White and red: the NO.08 Klesan (Men's Sleeveless Long Dress), the NO.14 Klesan (Men's Sleeveless Short Dress), and the NO.69 Truku (Women's Long Skirt).

White, red, and green: the NO.70 Truku (Unisex Shawl).

White, red, and blue: the NO.71, NO.73 Truku (Unisex Shawl).

Excluding textiles with black and grey as the primary color scheme, it is found that there are more textiles with white and red as the primary color scheme, mainly in the Klesan and Truku.

(4) Pairings of the red primary color scheme.

Red, Orange: the NO.39 Bunsui (Women's Long Sleeve Long Dress).

Red, Blue: the NO.16 Klesan (Men's Long Sleeve Short Dress), the NO.23 Klesan (Unisex Shawl), and the NO.25 Malepa (Men's Chest Cover).

Excluding textiles with black, grey, and white as the primary color scheme, it is found that there are more textiles with red and blue as the primary color scheme, mainly in Klesan and Malibu.

(5) Pairings of the yellow primary color scheme.

Yellow, Blue: the NO.19 Klesan (Women's Belt), the NO.20 Klesan (Women's Foot Protector), the NO.50 Beishi (Unisex Shawl).

Currently, only textiles from Klesan and Beishi are found to have yellow and blue as the primary color scheme for the research samples.

A preliminary understanding of the Atayal sub-group costumes has been obtained through the analysis and summary of the overall color use of the above textiles. However, detailed observation and analysis are still necessary. To organize the research samples, 367 sets of compound stripe colors can be obtained, with a total of 1033 color occurrences available for research and analysis, as shown in Appendix 4. Through descriptive statistics of color occurrences, it can be found that the black series has 196 occurrences, accounting for 19% of the total; the grey series has 173 occurrences, accounting for 16.7% of the total; the white series has 160 occurrences, accounting for 15.5% of the total; the red series has 187 occurrences, accounting for 18.1% of the total; the blue series has 127 occurrences, accounting for 12.3% of the total. These color series are the ones with more usage in compound stripe colors, distributed in various sub-groups of the research samples, with the blue series mainly concentrated in the Klesan, Malepa, Taranan, Plnawan, Bunsui, Beishi, Nanshi, Meicyubus, and Truku. The next are the ones with less usage in compound stripe colors, such as the peach red series has 24 occurrences, accounting for 2.3% of the total; the pink series has 26 occurrences, accounting for 2.5% of the total; the orange series has 27 occurrences, accounting for 2.6% of the total; the yellow series has 34 occurrences, accounting for 3.3% of the total; the green series has 38 occurrences, accounting for 3.7% of the total; the purple series has 15 occurrences, accounting for 1.5% of the total; the brown series has 26 occurrences, accounting for 2.5% of the total. Among them, the purple series has the fewest occurrences, only in NO.22 Klesan (Unisex Shawl), NO.40 Bunsui (Women's Long Sleeve Long Dress), NO.47 Beishi (Women's Long Sleeve Short Dress), NO.48 Beishi (Women's Foot Protector), NO.68 Truku (Women's Long Skirt), NO.71 Truku (Unisex Shawl).

Combining more than two colors to create a new visual effect can be called color matching. The color composition and configuration rules of Atayal tribal textiles have always been vague. This study attempts to organize the configuration methods of pattern colors from the research samples into a color-matching system for future practical application. The color combinations of classified compound stripes can be divided into two, three, four, five, six, and seven. Observing the performance of pattern-matching colors can more precisely understand the degree of preference for the use of color configuration by the Atayal tribe, as shown in Table 5.4. As follows:

(1) Two-color combination: The compound stripe colors are primarily two-color combinations, totaling 184 groups. Black (28.8%), grey (19.8%), white (24.7%), and red (16.3%) are the colors with higher usage rates, distributed in NO.01 Hakul (Hakku); NO.02 Makanji; NO.03, NO.04 Msbtunux; NO.05, NO.06, NO.07, NO.08, NO.09, NO.10, NO.11, NO.12, NO.14, NO.16, NO.20, NO.23 Klesan; NO.24, NO.26 Malepa; NO.27, NO.28, NO.29 Taranan; NO.30 Plnawan; NO.31, NO.32, NO.33, NO.34, NO.35, NO.36, NO.37, NO.39 Bunsui; NO.42, NO.44, NO.45 Beishi; NO.51, NO.52, NO.53 Nanshi; NO.56 Meicyubus; NO.58, NO.60, NO.61, NO.62 Musha; NO.63 Pulibau; NO.64, NO.66 Truku.

    Color matching performance: black, white (59 groups); black, grey (33 groups); white, red (24 groups); grey, red (20 groups); grey, blue (15 groups); black, red (10 groups); white, blue (7 groups); red, orange (5 groups); grey, peach red (3 groups); black, yellow (2 groups); black, blue (2 groups); grey, brown (2 groups); white, orange (1 group); red, blue (1 group), a total of 14 kinds of color matching combinations, as shown in Fig. 5.37.

(2) Three-color combination: The compound stripe colors are the second most common in three-color combinations, totaling 124 groups. Black (16.4%), grey (21.8%), red (23.4%), blue (16.9%) are the colors with higher usage rates, distributed in NO.03 Msbtunux; NO.05, NO.06, NO.08, NO.11, NO.12, NO.13, NO.14, NO.15, NO.16, NO.17, NO.18, NO.20, NO.21 Klesan; NO.24, NO.25, NO.26 Malepa; NO.30 Plnawan; NO.31, NO.32, NO.34, NO.35, NO.36, NO.37, NO.39 Bunsui; NO.44, NO.45, NO.48, NO.49, NO.50 Beishi; NO.51, NO.53,

Table 5.4 Analysis of color combinations of composite stripes in Atayal fabrics

Number of colors		Composite stripe color combination												Quantity Subtotal	Pattern Subtotal
		No color			Colorful										
		Black	Gray	White	Red	Peach pink	Pink	Orange	Yellow	Green	Blue	Purple	Brown		
Two colors	Occurrence	106	73	91	60	3	0.0	6	2	0.0	25	0.0	2	368	184
	Ratio (%)	28.8	19.8	24.7	16.3	0.8	0.0	1.6	0.5	0.0	6.8	0.0	0.5	100	
Three colors	Occurrence	61	81	35	87	6	3	4	4	9	63	3	16	372	124
	Ratio (%)	16.4	21.8	9.4	23.4	1.6	0.8	1.1	1.1	2.4	16.9	0.8	4.3	100	
Four colors	Occurrence	11	8	13	13	11	5	3	8	13	21	4	2	112	28
	Ratio (%)	9.8	7.1	11.6	11.6	9.8	4.5	2.7	7.1	11.6	18.8	3.6	1.8	100	
Five colors	Occurrence	4	4	8	11	1	4	3	7	8	8	4	3	65	13
	Ratio (%)	6.2	6.2	12.3	16.9	1.5	6.2	4.6	10.8	12.3	12.3	6.2	4.6	100	
Six colors	Occurrence	7	2	8	10	1	6	7	6	5	4	3	1	60	10
	Ratio (%)	11.7	3.3	13.3	16.7	1.7	10.0	11.7	10.0	8.3	6.7	5.0	1.7	100	
Seven colors	Occurrence	7	5	5	6	2	8	4	7	3	6	1	2	56	8
	Ratio (%)	12.5	8.9	8.9	10.7	3.6	14.3	7.1	12.5	5.4	10.7	1.8	3.6	100	
Sum	Occurrence	196	173	160	187	24	26	27	34	38	127	15	26	1033	367
	Ratio (%)	19.0	16.7	15.5	18.1	2.3	2.5	2.6	3.3	3.7	12.3	1.5	2.5	100	

Description: Atayal textiles' color composition and configuration rules have always been vague. This study attempts to organize the configuration and composition of pattern colors into a color-matching system from the research samples, which can be used as a reference for future practical applications. The color combinations of classified composite stripes can be divided into two-color, three-color, four-color, five-color, six-color, and seven-color combinations. By observing the performance of pattern color matching, we can more precisely understand the Atayal people's preference for color configuration (Drawn by this study)

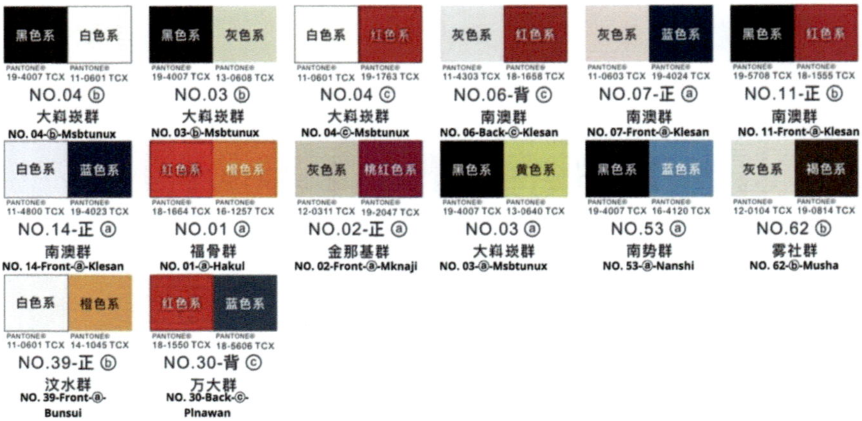

Fig. 5.37 Compound stripe two-color combination (sorted by decreasing repetition quantity) (drawn by this study) (Description: The most common composite stripe colors are two-color combinations, totaling 184 groups. Black (28.8%), gray (19.8%), white (24.7%), and red (16.3%) are the colors with the highest usage rates)

NO.54 Nanshi; NO.55, NO.56 Meicyubus; NO.57, NO.59, NO.60, NO.62 Musha; NO.65, NO.67, NO.68, NO.69, NO.72, NO.73 Truku.

Color representation: Grey, Red, Blue (39 sets); Black, Grey, Red (18 sets); Black, Grey, Brown (15 sets); White, Red, Blue (13 sets); Black, White, Red (11 sets); Black, Grey, Blue (4 sets); White, Red, Green (3 sets); White, Green, Purple (3 sets); Black, Peach, Orange (2 sets); Black, Grey, Peach (2 sets); Grey, Yellow, Blue (2 sets); Black, White, Orange (1 set); Black, White, Green (1 set); Black, White, Blue (1 set); Black, Red, Green (1 set); Black, Red, Blue (1 set); Black, Peach, Blue (1 set); Black, Pink, Yellow (1 set); Black, Yellow, Brown (1 set); Black, Green, Blue (1 set); Grey, Red, Peach (1 set); White, Pink, Orange (1 set); White, Pink, Blue (1 set), totaling 23 color combinations, as shown in Fig. 5.38.

(3) Four-color combinations: Only 28 sets of composite stripes colors with four-color combinations exist. White (11.6%), Red (11.6%), Green (11.6%), Blue (18.8%) are the colors with higher usage rates, distributed in NO.10, NO.19, NO.21, NO.22 Klesan; NO.34, NO.35, NO.36, NO.38 Bunsui; NO.45, NO.47, NO.49, NO.50 Beishi; NO.59 Musha; NO.65, NO.69, NO.70 Truku.

Color representation: White, Red, Pink, Blue (4 sets); Grey, Peach, Green, Blue (4 sets); Black, Peach, Blue, Purple (3 sets); Grey, Red, Green, Blue (3 sets); Black, White, Red, Peach (1 set); Black, White, Red, Yellow (1 set); Black, White, Red, Green (1 set); Black, White, Red, Blue (1 set); Black, Grey, Yellow, Brown (1 set); Black, Peach, Yellow, Green (1 set); Black, Peach, Yellow, Blue (1 set); Black, Orange, Green, Blue (1 set); White, Red, Yellow, Green (1 set); White, Red, Green, Blue (1 set); White, Peach, Orange, Blue (1 set); White,

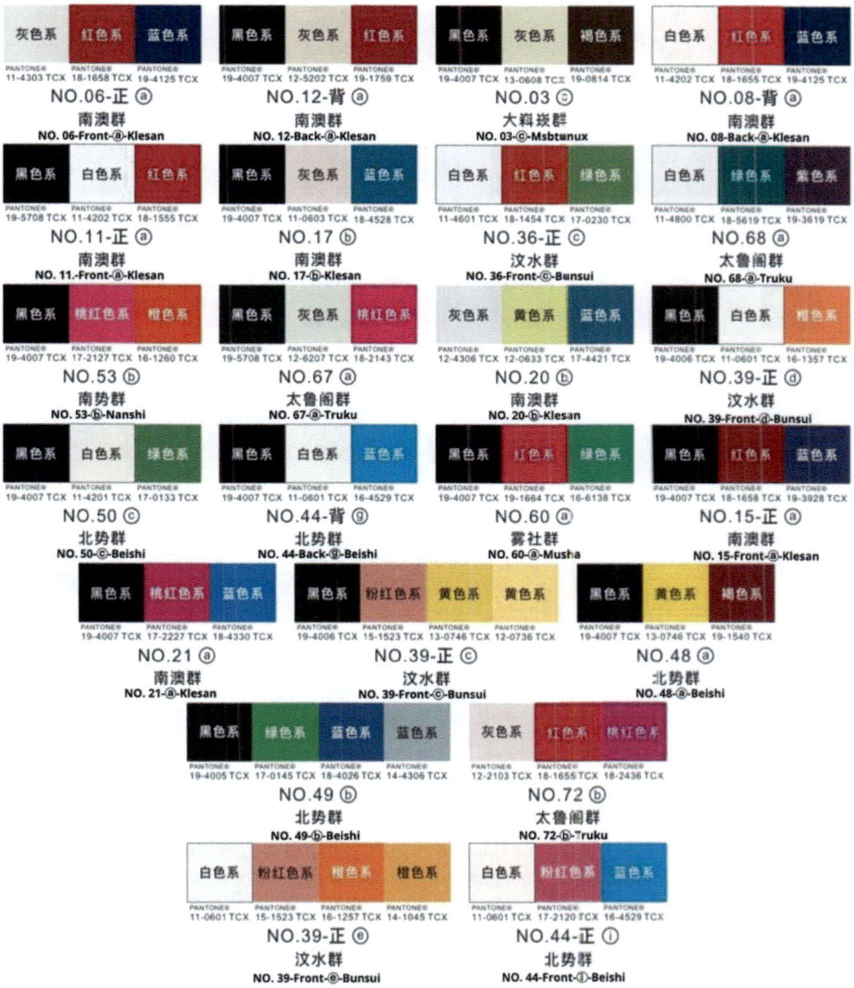

Fig. 5.38 Composite stripe three-color combinations (sorted in descending order of repetition) (Drawn by this study) (Description: The most common composite stripe colors are three-color combinations, totaling 124 groups. Black (16.4%), gray (21.8%), red (23.4%), and blue (16.9%) are the most used colors)

Pink, Yellow, Blue (1 set); White, Yellow, Green, Blue (1 set); Orange, Yellow, Purple, Brown (1 set), totaling 18 color combinations, as shown in Fig. 5.39.

(4) Five-color combinations: There are only 13 sets of composite stripes colors with five-color combinations. White (12.3%), Red (16.9%), Yellow (10.8%), Green (12.3%), Blue (12.3%) are the colors with higher usage rates, distributed in NO.36, NO.39, NO.40 Bunsui; NO.41, NO.44, NO.47, NO.48, NO.50 Beishi; NO.51 Nanshi; NO.71 Truku.

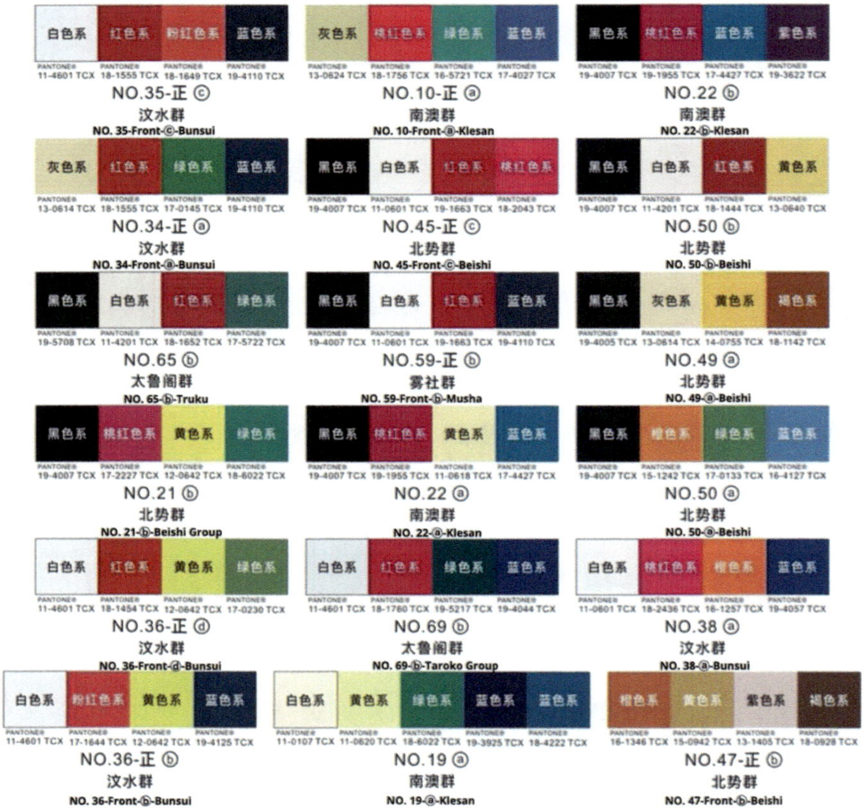

Fig. 5.39 Composite stripe four-color combinations (sorted in descending order of repetition) (Drawn by this study) (Description: There are only 28 sets of four-color combinations of composite stripes. White (11.6%), red (11.6%), green (11.6%), and blue (18.8%) are the most used colors)

Color representation: Grey, Red, Green, Blue, Purple (2 sets); Black, White, Red, Pink, Green (1 set); Black, White, Red, Green, Blue (1 set); Black, White, Green, Blue, Purple (1 set); Black, Red, Peach, Orange, Yellow (1 set); Grey, White, Red, Pink, Yellow (1 set); Grey, Red, Yellow, Green, Blue (1 set); White, Red, Pink, Orange, Yellow (1 set); White, Red, Yellow, Green, Blue (1 set); White, Red, Yellow, Purple, Brown (1 set); White, Red, Green, Blue, Brown (1 set); Pink, Orange, Yellow, Blue, Brown (1 set), totaling 12 color combinations, as shown in Fig. 5.40.

(5) Six-color combination: There are only ten sets of composite stripes colors with six-color combinations. Black (11.7%), White (13.3%), Red (16.7%), Pink (10%), Orange (11.7%), and Yellow (10%) are the colors with higher usage rates, distributed in NO.39, NO.40 Bunsui; NO.41, NO.46, NO.47 Beishi; NO.71 Truku.

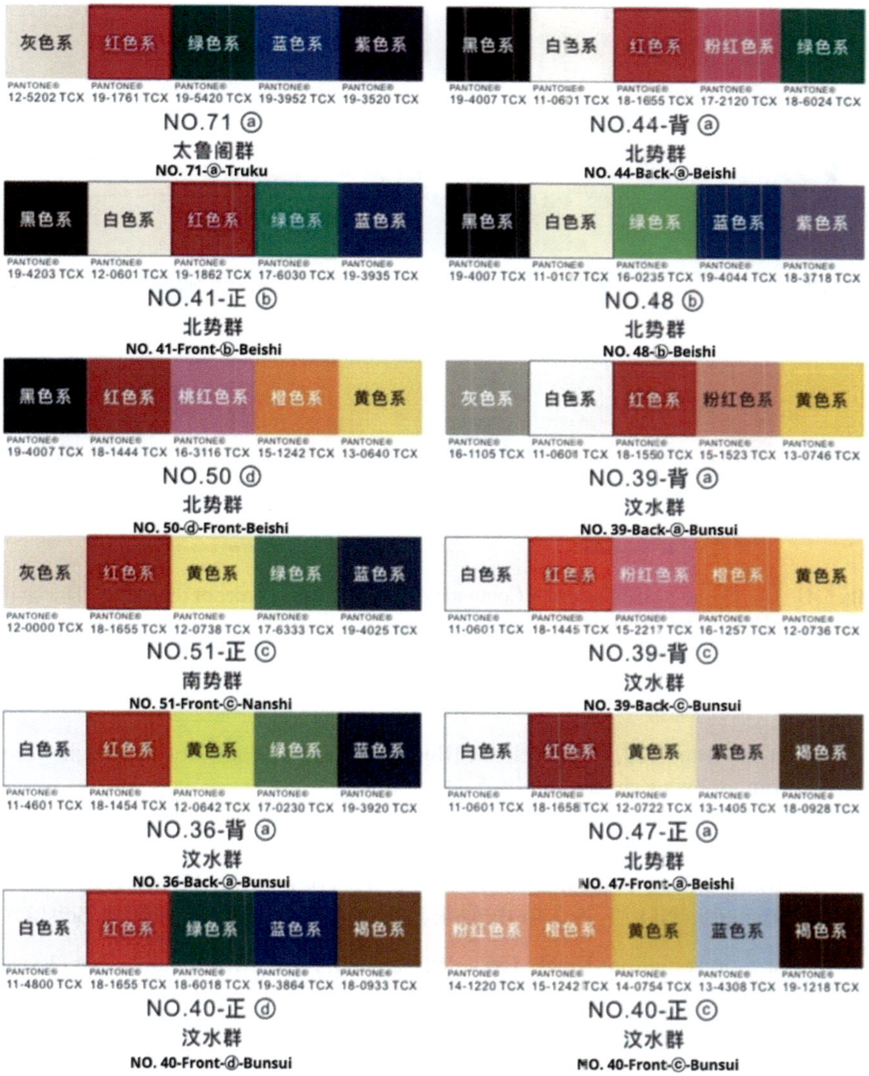

Fig. 5.40 Composite Stripe Five-Color Combination (in descending order of repetition) (Drawn by this study) (Description: There are only 13 combinations of five colors for composite stripes. White (12.3%), red (16.9%), yellow (10.8%), green (12.3%), and blue (12.3%) are the most used colors)

Color representation: Black, White, Red, Pink, Orange, and Yellow (2 sets); Black, White, Red, Orange, Yellow, and Purple (2 sets); Black, Grey, Red, Pink, Yellow, and Green (1 set); Black, White, Red, Pink, Orange, and Green (1 set); Black, White, Red, Orange, Green, and Blue (1 set); Grey, Red, Peach, Green, Blue, and Purple (1 set); White, Red, Pink Orange, Green, and Blue

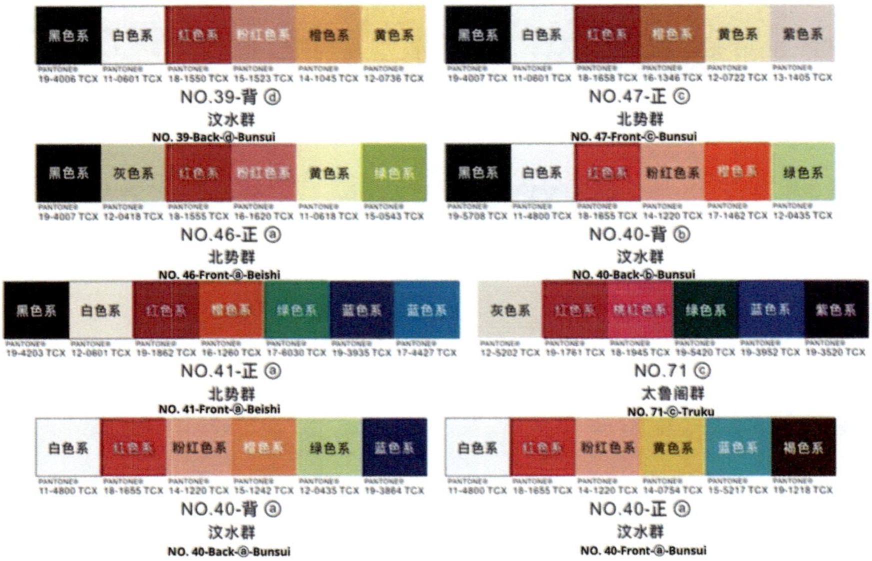

Fig. 5.41 Composite Stripe Six-Color Combination (in descending order of repetition) (Drawn by this study) (Description: There are only 10 combinations of six colors for composite stripes. Black (11.7%), white (13.3%), red (16.7%), pink (10%), orange (11.7%), and yellow (10%) are the most used colors)

(1 set); White, Red, Pink, Yellow, Blue, and Brown (1 set), totaling eight color combinations, as shown in Fig. 5.41.

(6) Seven-color combination: There are only eight sets of composite stripe colors with seven-color combinations. Black (12.5%), Red (10.7%), Pink (14.3%), Yellow (12.5%), and Blue (10.7%) are the colors with higher usage rates, distributed in NO.39, NO.40 Bunsui; NO.42, NO.43, NO.44, NO.46 Beishi.

Color representation: Black, Grey, White, Red, Pink, Orange, and Blue (1 set); Black, Grey, Red, Pink, Orange, Yellow, and Blue (1 set); Black, Grey, Red, Pink, Yellow, Green, and Blue (1 set); Black, Grey, Peach, Pink, Yellow, Green, and Blue (1 set); Black, White, Red, Peach, Pink, Yellow, and Blue (1 set); Black, White, Red, Pink, Orange, Yellow, and Brown (1 set); Black, White, Red, Pink, Yellow, and Green, Purple (1 set); Grey, White, Pink, Orange, Yellow, Blue, and Brown, totaling eight color combinations, as shown in Fig. 5.42.

The colors of the Atayal tribe's composite stripes can be summarized into 12 color systems. The grouping and analysis of the color mentioned above combinations found that black, grey, white, red, and blue are the most frequently used in composite stripes, followed by peach red, pink, orange, yellow, green, purple, and brown. These different combinations of colors clearly show a difference in quantity. In two-color combinations, the quantities of black, grey, and white are higher than other colors, with red being the most numerous among the colored ones. In three-color combinations, the quantities of black, grey, and white decrease, while the quantities of red and

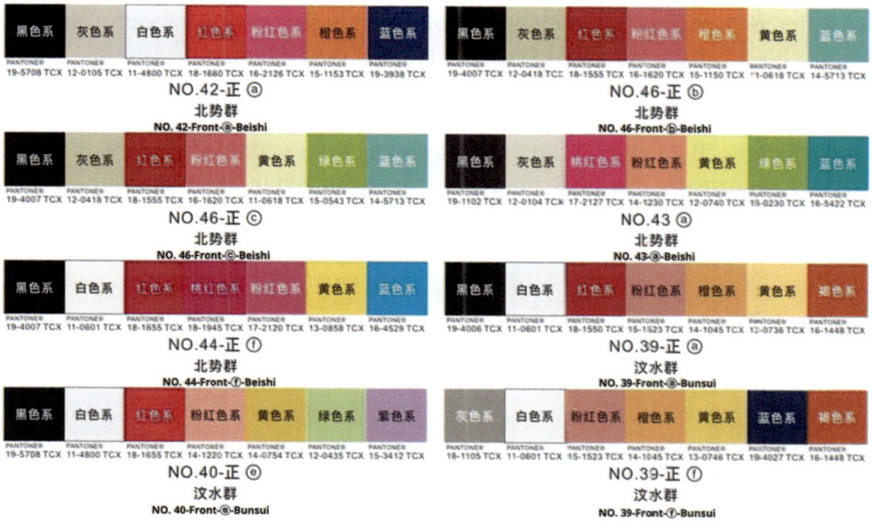

Fig. 5.42 Composite Stripe Seven-Color Combination (Drawn by this study) (Description: There are only eight combinations of seven colors for composite stripes. Black (12.5%), red (10.7%), pink (14.3%), yellow (12.5%), and blue (10.7%) are the most used colors)

blue increase significantly. The red, green, and blue quantities surpass black, grey, and white in four-color and five-color combinations. In six-color combinations, the amount of red is higher than other colors, while the quantities of pink, orange, and yellow increase. In seven-color combinations, the quantities of black, grey, white, red, and blue are slightly different, while the amounts of pink and yellow continue to increase. Although colorless black, grey, and white are the primary colors used in composite stripes, other colors become the primary colors in four-color, five-color, six-color, and seven-color combinations.

The composite stripes of the Atayal tribe's textiles use the most multi-color combinations. These color combination methods use colorless black, grey, white, and other colored hues. The overall effect is due to the corresponding chroma forming a rich color appearance. In addition, there are 14 kinds of two-color combinations, 23 types of three-color combinations, 18 kinds of four-color combinations, 12 kinds of five-color combinations, eight kinds of six-color combinations, and eight kinds of seven-color combinations. Two-color and three-color combinations have the most coloring styles and appear most frequently. In contrast, four-color, five-color, six-color, and seven-color combinations show a decreasing number of color combinations, with fewer repetitive color combinations, and even seven-color composite stripes only appear in the Bunsui and Beishi's textiles. For color combinations, such as NO.39-Positive© Bunsui's color combination is black, pink, and yellow, but yellow has two different depths of chroma. Similar color configurations also appear in NO.49ⓑ Beishi's black, green, blue (dark blue, light blue); NO.39-Positiveⓔ Bunsui's white, pink, orange (dark orange, light orange); NO.19ⓐ Klesan's white, yellow, green, blue (dark blue, light blue); NO.41-Positiveⓐ Beishi's black, white, red, orange, green, blue (dark blue, light blue). From this, the Atayal tribe's composite stripes have the most comprehensive distribution range for two-color and three-color combinations, and the coloring styles are the most abundant.

Chapter 6
Pan-Atayal Group Textile

6.1 Logic of Textile Pattern Form

The Atayal tribe uses the characteristics of woven patterns to create many different forms of patterns and various color combinations in clothing. What seem to be ordinary geometric patterns hide rich forms, undoubtedly demonstrating great weaving ability. Looking at the pattern types, primary forms, and compound stripe structures of Atayal textiles, it can be inferred that there are no two pieces of clothing with the same pattern. Even similar patterns have some subtle differences, and many of the same primary forms of patterns can be seen in different sub-group textiles. At the same time, there is a systematic way of constructing textile patterns and applying modular thinking when weaving patterns can enhance the expression of weaving art, thereby increasing various forms of woven patterns. The diamond, mountain, and cross shapes of Atayal textile patterns, commonly seen in traditional clothing, should be inevitable phenomena in weaving. However, due to critical factors such as the range, level, or color matching of the woven patterns and the evolution over a long weaving period, these pattern forms have been intentionally or unintentionally created in complex or simplified ways, creating almost formulaic pattern rules. Today, these woven patterns are quite ethnic in style. However, the regularization of design thinking leads to a specific pattern of textile patterns, which has become the Atayal people's understanding of the meaning of form. The inheritance of form and collective consciousness determine the continuity of the creation of woven patterns and form the unique artistic connotation of the nation from the expression of its pattern types, primary forms, and so on.

Systematic modules can extend many forms and build patterns into a database, especially for Indigenous peoples who only have oral transmission without written records; it is essential! The following case studies focus on the module systems of specific textiles, discussing the design thinking of Atayal weaving patterns. Here, we choose the same type of textile and a sub-group with a more significant number of patterns as the research object, so we use the NO.22 Klesan of the Tayal and the

P. H. Wang and J. Li, *Interpreting the Evolution of Atayal Tribal Textile Patterns*,
https://doi.org/10.1007/978-981-96-5451-2_6

NO.61 Musha and NO.70 Truku of the Seediq to infer the formal logic of pattern construction.

(1) NO.22 Men's and women's shawls are Klesan woven products; the entire garment is filled with horizontal stripes, its composite stripes are composed of diamond patterns, and the whole looks like it is composed of repeated patterns of the same shape. There are still many differences. The inner layer of the pattern is composed of 14 basic shapes, which can be classified into diamonds, mountains, crosses, × shapes, and other basic shapes. By analyzing the structure in the middle position of the pattern, the 35 patterns of NO.22 can be summarized into nine different types of modules, as shown in Fig. 6.1. Each type has a common feature of a ⟨⟩ frame; the difference lies in the different organizational structures within the frame, mainly using × and ◇ two shapes for combination changes. The description of the pattern formation rules is as follows:

   (a) × shape is the core structure, as shown in Fig. 6.1ⓐ.
   (b) A structure like × appears in the upper and lower positions in V and inverted V, as shown in Fig. 6.1ⓑ.
   (c) The core structure appears in the upper and lower positions in W and inverted W, as shown in Fig. 6.1ⓒ.
   (d) The core structure appears in the upper and lower positions in M and inverted M, as shown in Fig. 6.1ⓓ.
   (e) Like a grid, the core structure is shown in Fig. 6.1ⓔ.
   (f) The structure intersects into a diamond shape with⟩ and ⟨shapes, as shown in Fig. 6.1ⓕ.
   (g) The diamond structure extends diagonally from the upper and lower edges, as shown in Fig. 6.1ⓖ.
   (h) The diamond is the core structure, as shown in Fig. 6.1ⓗ.
   (i) The diamond is structured in groups, as shown in Fig. 6.1ⓘ.

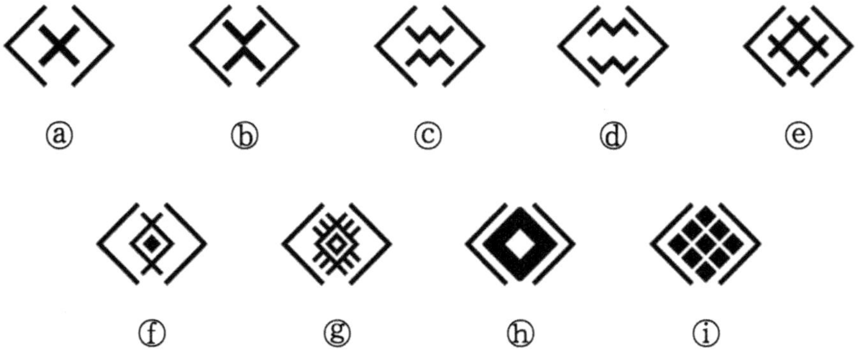

Fig. 6.1 Rule inference of pattern systematization module based on NO.22 men's and women's shawls (drawn by this study)

(2) NO.61 men's and women's shawls, the entire fabric is filled with horizontal compound stripes; its pattern feature is one of the typical styles of the Musha community textiles. Tsai Yu-shan also mentioned this weaving effect is also known as rice grain weaving [1]. This style of fabric, because all the weaves are the same color as if burying the pattern in the fabric, that is, weaving the pattern into the fabric during the weaving process, so that the texture is like the pattern, also causing difficulty in identifying the weave structure, therefore it must be observed from the back of the fabric. The inner layer of the fabric pattern is composed of 11 basic shapes, which can be classified as diamond, × shape, diagonal line shape, and other basic shapes. Deconstructing the core organization of the pattern, five different types of modules can be summarized from the eight patterns of NO.61, as shown in Fig. 6.2. Each module type is based on a diamond shape, with variations in the structure within the diamond. The description of the pattern formation rules is as follows:

(a) × shape is the core structure, as shown in Fig. 6.2ⓐ.
(b) The × shape structure comprises⟩ and ⟨ shapes, as shown in Fig. 6.2ⓑ.
(c) The hashtag shape is the core structure, as Fig. 6.2ⓒ shows.
(d) Large and small diamonds overlap and are aligned to the left, as Fig. 6.2ⓓ shows.
(e) The large and small diamonds overlap aligned to the right, as shown in Fig. 6.2ⓔ.

(3) The NO.70 men's and women's shawls are filled with diamond patterns, the most typical style of the Truku textiles. Overall, it seems composed of repeated patterns of the same shape, but individual differences remain. The inner layer of the pattern is composed of 17 basic shapes, which can be classified as diamond, cross, and × shapes. Based on the diamond structure in the middle of the pattern as the analysis principle, nine different modules can be summarized from the 57 patterns of NO.70, as shown in Fig. 6.3. Each module type has a "◆◆" shape feature, and the difference lies in the structures inside the frame. The description of the pattern formation rules is as follows:

(a) The × shape is the core structure, as shown in Fig. 6.3ⓐ.
(b) The diamond overlaps with the × shape, as shown in Fig. 6.3ⓑ.
(c) The diamond overlaps with the thicker × shape, as shown in Fig. 6.3ⓒ.

Fig. 6.2 Rule inference of the systematic module of pattern NO 61 men's and women's shawls (drawn by this study)

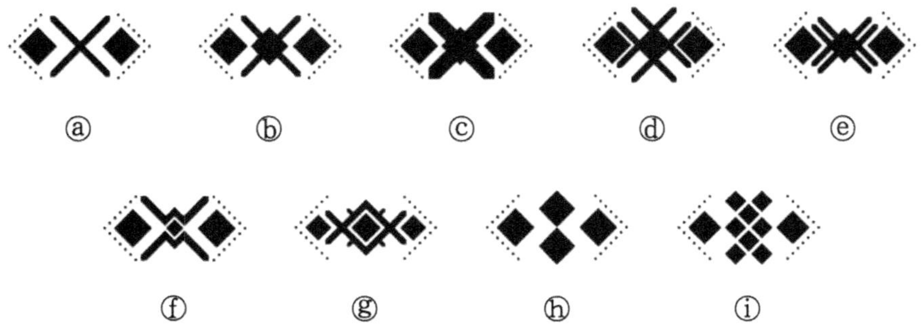

Fig. 6.3 Rule inference of the pattern systematization module based on NO.70 male and female shawls (drawn by this study)

(d) The diamond combines with the⟩ and ⟨ shapes at the top, bottom, left, and right, as shown in Fig. 6.3ⓓ.
(e) The diamond combines with the ⟫ and ⟪ shapes on the left and right sides, as shown in Fig. 6.3ⓔ.

The diamond combines with the W and inverted W shape at the top and bottom, as shown in Figure 6.3ⓕ.

The diamond overlaps with the ⟩◇⟨ shape, as shown in Figure 6.3ⓖ.

The diamond uses a vertical repetition as its structure, as shown in Figure 6.3ⓗ.

The diamond uses a vertical grouping as its structure, as shown in Figure 6.3ⓘ.

Ethnic style is a unique artistic characteristic formed by a group over a long development period, constituted by factors such as a nation's social structure, customs, religious beliefs, etc. Ethnic style is a country's unique cultural symbol or characteristic, an "ethnic element." It can be explained that the module contains several factors by summarizing the above examination of modules in the weaving structure, i. The module is a replaceable component, combined into patterns according to rules and inertia, and its form inevitably shows subtle similarities or differences, such as the pattern modules in Figs. 6.1ⓐ, 6.2ⓐ, and 6.3ⓐ. This × structure feature also appears in other textiles, such as the NO.10-BackⓓⓕKlesan (men's sleeveless long clothes), the NO.60ⓒMusha (men's and women's shawls), the NO.68ⓐTruku (women's long skirts), etc. Furthermore, in the pattern module in Fig. 6.1ⓖ, this structural feature can also be seen in other sub-group textiles, such as the NO.03ⓒⓓMsbtunux (men and women's shawls), the NO.10-BackⓘKlesan (men's sleeveless long clothes), the NO.39-FrontⓑBunsui (women's long-sleeved long clothes), etc. From the comparison and analysis of these textiles, it can be inferred that the Atayal tribe's continuation of weaving skills has explored the regularity of weaving patterns through the method of pattern shaping module, which is undoubtedly a clear example.

In summary, if you master the pattern systematization module method during weaving, you can create many new patterns. The main principle of creating new

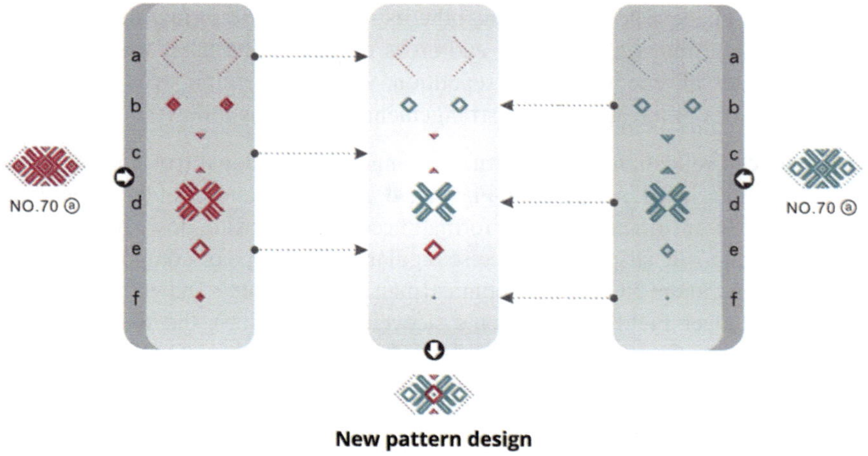

Fig. 6.4 Deconstruction and reconstruction of the pattern systematization module (drawn by this study)

combinations is module interchange, so these module themes can appear in various variant forms and can be combined into highly complex patterns. According to the above definition, the module is a replaceable component. Here, using the deconstruction and reconstruction of the pattern systematization module, redesign a pattern, such as the design example in Fig. 6.4, and select two patterns from the 57 patterns of the NO.70 Truku (men and women's shawls), each taking the constituent elements in the pattern, the component organization has up and down or left and proper symmetry, as well as the single diamond component of the pattern core, etc., and then recombine to form a new pattern, this mode can also create other shapes. There are different levels of complexity in the pattern module. Complex modules can contain simpler other modules, such as repeating some simple patterns, which can also change shapes. When these themes appear in endless combinations, they can also change shapes to adapt to their specific positions in the weaving structure.

6.2 Composite Stripe Arrangement of Textiles

The patterns of Atayal textiles are mainly composed of geometric themes; the interpretation of geometry is full of imagination, order, and rigor and has a strong sense of rationality. This study has discussed the structure of composite stripes in the fifth chapter, thus understanding the organizational performance of textile patterns. Here, the analysis of the composite stripes of Atayal textiles, the repetitive characteristics of its unit patterns, indeed have obvious arrangement rules. Through the repetitive structure of composite stripes, two types of patterns shaping performances can be summarized: one is reflection, and the other is repetition; the former includes staggered reflection, interval reflection, and cross reflection; the latter has translation repetition, interval repetition, cross repetition, vertical repetition, group repetition. The descriptions of the above eight arrangement rules are as follows:

(1) Interlaced reflection: The pattern is arranged upside-down (rotated 180°) and closely interlaced. As shown in Fig. 6.5, ❶ and ❷ are patterns of the same shape but in different directions. Their formation comprises upside-down reflections to form composite stripes with precise regularity. This type of composite stripe is also found in the NO.03ⓐ Msbtunux (men's and women's shawls), the NO.56. Frontⓐ Deer Field Group (men's sleeveless long robe), the NO.58-Frontⓔ Fog Society Group (men's long-sleeved long robe), the NO.58-Backⓓ Fog Society Group (men's long-sleeved long robe), the NO.66ⓐ Truku (women's set sleeves), etc.
(2) Interval reflection: The pattern is arranged upside-down horizontally, with equal intervals between the patterns or regularly interspersed with another group of patterns. As shown in Fig. 6.6, ❶ and ❷ are patterns of the same shape but in different directions. Their formation is composed of upside-down reflections to form composite stripes. This type of composite stripe is also found in the NO.03ⓑ Msbtunux (men's and women's shawls), the NO.06. Frontⓕ Klesan (men's sleeveless long robe), the NO.21ⓐ Klesan (men's and women's shawls), the NO.63-Backⓗ Pulibau (men's sleeveless long robe), etc.
(3) Crossing reflection: The pattern is arranged upside-down and vertically symmetrical, with equal intervals between the patterns. There is a gap or a band of stripes in the middle position of the upper and lower rows. As shown in Fig. 6.7, ❶ and ❷ are patterns of the same shape but in different directions. Their formation is composed of upside-down reflections to form composite stripes. This type of composite stripe is also found in the NO.30-Frontⓐ Plnawan (men's

Fig. 6.5 Alternating reflection (drawn by this study)

Fig. 6.6 Interval reflection (drawn by this study)

long-sleeved long robe), the NO.50ⓑ North Power Group (men's and women's shawls), the NO.53ⓓ South Power Group (women's long skirt), etc.

(4) Translation repetition: The pattern is arranged in a horizontal repetition, as shown in Fig. 6.8; ❶ and ❷ are patterns of the same shape, forming composite stripes with the same shape. This type of composite stripe is also found in the NO.06. Backⓔ Klesan (men's sleeveless long robe), the NO.23ⓑ Klesan (men's and women's shawls), the NO.26ⓒ Malepa (tobacco bag), the NO.34-Frontⓔ Bunsui (men's and women's long-sleeved long robe), the NO.44-Frontⓓ North Power Group (women's long-sleeved long robe), the NO.55-Frontⓑ Deer Field Group (men's sleeveless long robe), etc.

(5) Interval repetition: The pattern is arranged horizontally, with a gap between the two or interspersed with another group of patterns. As shown in Fig. 6.9, ❶ and ❷ are patterns of the same shape, forming composite stripes with the same shape. This type of composite stripe is also found in the NO.07-Frontⓐ Klesan (men's sleeveless long robe), NO.10-Backⓒ Klesan (men's sleeveless long robe), the NO.14-Frontⓒ Klesan (men's sleeveless short robe), the NO.28ⓐ Taranan (women's long skirt), NO.37-Frontⓔ Bunsui (men's long-sleeved short robe), the NO.58-Backⓖ Fog Society Group (men's long-sleeved long robe), etc.

(6) Overlapping repetition: The pattern repeats vertically, arranged symmetrically, with equal spacing between the patterns. There is a gap or a band of stripes in the middle of the upper and lower rows, as shown in Fig. 6.10, where ❶ and ❷ are patterns of the same shape, composed of the same form to form a composite

Fig. 6.7 Overlapping reflection (drawn by this study)

Fig. 6.8 Translation repetition (drawn by this study)

Fig. 6.9 Interval repetition (drawn by this study)

stripe. This type of composite stripe is also found in NO.05-Front ⓐ Klesan (Men's Sleeveless Long Robe), the NO.06. Front ⓑ Klesan (Men's Sleeveless Long Robe), the NO.14-Front ⓑ Klesan (Men's Sleeveless Short Robe), the NO.16. Front ⓕ Klesan (Men's Long Sleeve Short Robe), the NO.35-Back ⓑ Bunsui (Men's Long Sleeve Long Robe), etc.

(7) Vertical repetition: The pattern is arranged in a vertical or interlaced repetition, as shown in Fig. 6.11, where ❶ and ❷ are patterns of the same shape, composed of the same form to form a composite stripe. This type of composite stripe is also found in NO.11-Front ⓑ Klesan (Men's Sleeveless Long Robe), the NO.33-Back ⓑ Bunsui (Men's Sleeveless Long Robe), the NO.38ⓐ Bunsui (Men's Chest Cover), the NO.57-Front ⓐⓑ Musha (Men's Sleeveless Long Robe), etc.

Fig. 6.10 Overlapping repetition

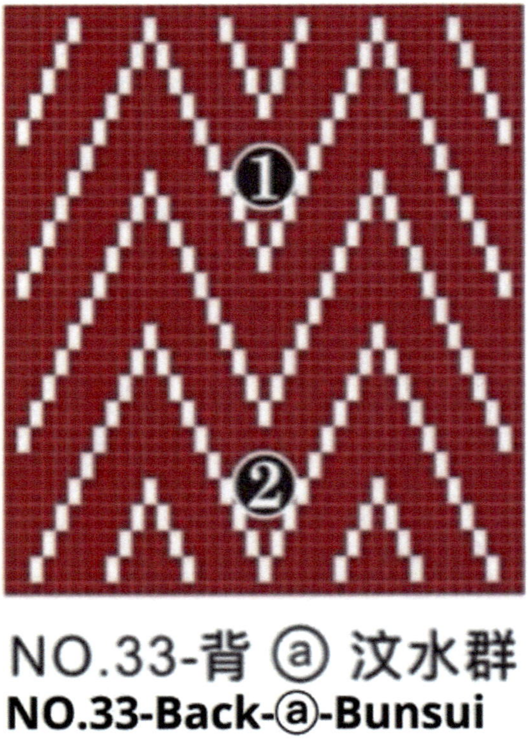

Fig. 6.11 Vertical repetition (drawn by this study)

Fig. 6.12 Grouped repetition (drawn by this study)

(8) Grouped repetition: The pattern is arranged in a vertical or horizontal repetition, forming a clustered shape, as shown in Fig. 6.12, where ❶ and ❷ are patterns of the same shape, composed of the same form to form a composite stripe. This type of composite stripe is also found in the NO.24-Back ⓑ Malepa (Men's Sleeveless Long Robe), the NO.25ⓐ Malepa (Men`s Chest Cover), the NO.52ⓐ Nan Shi Group (Men's Chest Cover), the NO.58-Front ⓒⓖ Musha (Men's Long Sleeve Long Robe), NO.64ⓐ Truku (Men's Chest Cover), etc.

The composite stripes of the Atayal tribe's textiles listed above are based on rules that form a band-like organization, using reflective, repetitive, and other symmetrical structures arranged into orderly composite stripes. From the horizontal or vertical arrangement of the composite stripes, it is easy to perceive the textile pattern, for example, a section of the same shape pattern, composed of equal spacing and repetitive form, where the unit pattern cyclically appears several times. Thus, it can be defined as a regular composite stripe. In addition, because the weaving area limits the size of the pattern, it ends at the edge of the textile or is a truncated unit pattern. At this time, it is impossible to judge whether the pattern continues to cycle, such as the NO.17ⓐ Klesan (Women's Belt), the NO.36. Front ⓒⓔ Bunsui (Men's Long Sleeve Long Robe), etc., are respectively small weaving patterns and interlaced two kinds of patterns, forming a section of composite stripes without unit pattern cycle, so it cannot be determined by regularity. Some composite stripes' repetitiveness is not easy to identify. It seems complicated to define accurately, such as the NO.01ⓐⓑ Hakul (Hakku) (Men's and Women's Shawl), the NO.22ⓑⓒⓓ Klesan (Men's and Women's Shawl), and the NO.61ⓐⓑⓒⓓⓔⓕⓖⓗ Musha (Men's and Women's Shawl), the NO.56. Back ⓐⓑ Meicyubus (Men's Sleeveless Long Robe), the NO.62ⓐⓑⓒ Musha (Woven Mat), the NO.70ⓐ Truku (Men's and Women's Shawl), etc., these composite stripes have the above arrangement rules, but there are differences between the patterns. Their forms are similar but not the same, forming several different patterns interlaced in the same section of weaving; this situation violates the principle of the same form to form a composite stripe. However, this is one of the highly variable features of the Atayal tribe's textile composite stripes.

In summary, the compound stripes of Atayal tribal textiles are generally arranged symmetrically, with only a few irregularities present, which may be unintentional or intentional. The former usually involves the accidental insertion of unit patterns, a row of unit patterns in a section of weaving that are incorrectly oriented, or a change in shape to adjust the last unit pattern in a row, such as the NO.04ⓐ Msbtunux (men's and women's shawls), the NO.21ⓐⓑ Klesan (men's and women's shawls), the NO.39-Backⓐ Bunsui (women's long-sleeved dresses), the NO.40-Frontⓑ Bunsui (women's long-sleeved dresses), NO.62ⓑ Musha (woven mats), etc. Strictly speaking, these irregular woven patterns have broken the symmetry rule of compound stripes. One can also perceive attempts to create pattern rules from the Atayal weaving craft. Still, due to the method, technique, or decorative equipment, the precise arrangement of weaving positions may have some discrepancies, such as the NO.54ⓐ Nanshi (women's foot protection cloth) arranged in beads, which is a typical situation. This pattern is strung with beads vertically and horizontally, interwoven into the fabric, so it is impossible to be regular, which is a situation that is expected to occur. Moreover, the handmade grinding of beads inherently has size differences. Stinging beads and interweaving them into the fabric will naturally cause misalignment, but this is unintentional.

6.3 Correlation of Textile Pattern Colors

The fibers of Atayal tribal textiles are mostly dyed using dyeing techniques, with a single fiber typically dyed in one color. The various colors in the textiles are all composed of different hues and areas, interwoven through differently configured warp and weft lines. Upon close observation of the weaving structure, the colors of the patterns may appear to float on the surface of the fibers, or similar colored patterns in two places may disappear at the boundary due to their adjacent relationship. They may also produce a shimmering effect due to the vibrant colors. From the perspective of modern color theory, the composite stripes of Atayal tribal textiles are multi-hued color matching or multi-hued contrast, generally with three or more hues being the most common. The color combination method belongs to "juxtaposition additive color mixing." this mixing process can also be referred to as additive color operation, like the visual effect of dot matrix graphics mixing. This color combination is due to the color points on the surface of the color sample being small or the lines being thin, and the colors placed together will mutually influence, spread, and blend (mix) into the adjacent colors, forming new mixed colors. The expression of juxtaposition color mixing cannot accurately predict the result of the color mixing, but there should be some relationship between the pattern and the color. Therefore, the correlation between patterns and colors can be understood by comparing the types, primary forms, and composite stripe structures of colors and patterns.

Exploring the preference for color combinations in Atayal tribal textiles, the color combinations of patterns in clothing that form composite stripes are more extensive, especially NO.10 Klesan (men's sleeveless long robes); NO.34, NO.35, NO.36 Bunsui (men's long-sleeved long robes); NO.39, NO.40 Bunsui (women's long-sleeved long robes); NO.41, NO.42 Beishi (men's long-sleeved long robes); NO.44, NO.45 Beishi (women's long-sleeved long robes); NO.46, NO.47 Beishi (women's long-sleeved short robes); NO.51 Nanshi (men's long-sleeved long robes). The color combinations of shawls and foot protectors are primarily in three or four colors, such as NO.03 Msbtunux (men's and women's shawls), NO.21, NO.22 Klesan (men's and women's shawls); NO.48, NO.49 Beishi (women's foot protectors); NO.50 Beishi (men's and women's shawls); NO.70, NO.71, NO.72, NO.73 Truku (men's and women's shawls). Comparing the colors and patterns, it can be found that the composite stripe structure of textiles uses the same color or several colors in combination, and the layout and arrangement of the pattern structure are composed of shapes and colors. Observing the color performance of the pattern shape boundary, if the boundary is a light color series, it gets darker towards the inside. For example, in the pattern of NO.39-Positiveⓐ Bunsui, the dark diamond basic shape is a black series. Its horizontal separation structure color matching method is the same, as shown in Fig. 6.13. Conversely, if the shape boundary is a dark color series, it gets lighter towards the inside, such as the NO.47-Positiveⓒ Beishi pattern. The light cross's basic shape is a yellow series, and its horizontal separation structure color matching is also the same pattern, as shown in Fig. 6.14.

The examples above of color configurations in pattern-shaping boundaries belong to composite stripes with more diverse and layered colors. However, this is not a fixed color matching rule, and there may also be similar color combinations between shaping boundaries, such as the pattern of NO.39-Frontⓔ Bunsui, where the dark cross shape is basically in the orange range, the overall diamond pattern is a lighter orange. The pink range borders the edge contour. Therefore, the colors of the composite stripes will present a similar halo effect, as shown in Fig. 6.15. The color performance of simple pattern-shaping boundaries is more commonly used in black, grey, white, and other color ranges to separate patterns. These colors are sometimes also the base color (background color) of the pattern, as shown in Fig. 6.16, the pattern of NO.10-Backⓐ Klesan, where the cross shape is basically in the blue

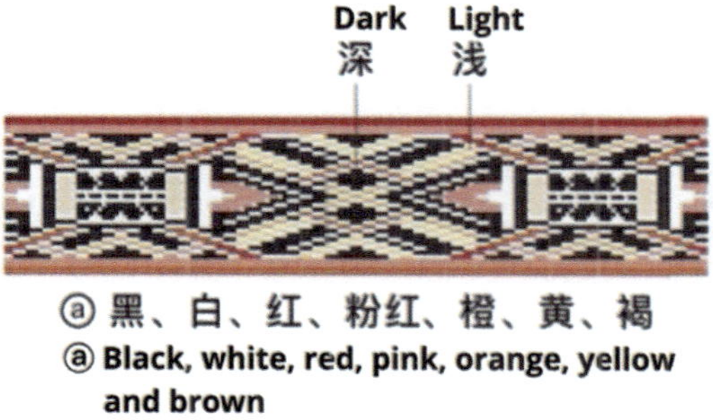

Fig. 6.13 Color performance of shape boundary with NO.39 as an example (drawn by this study)

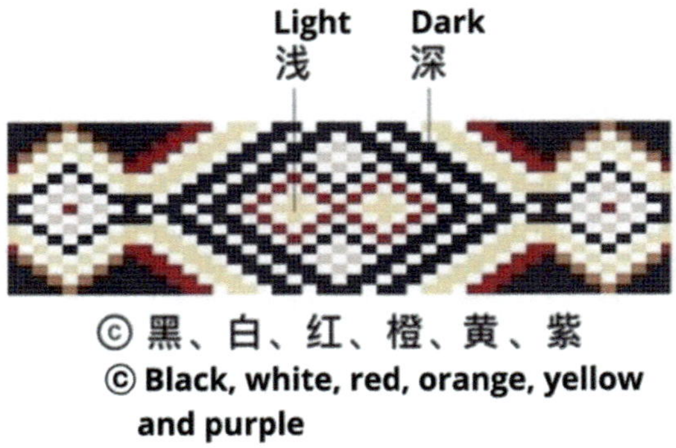

Fig. 6.14 Color performance of shape boundary with NO.47 as an example (drawn by this study)

range, the shaping boundary is in the grey range, and the structure is easy to separate the outer pattern. The color combination of its transverse repetitive structure is the same method, thus showing a repetitive color configuration. The color performance of pattern shaping, black, grey, white, and other achromatic colors can be said to be the primary colors of the pattern. In addition to serving as the boundary of the basic shape and woven structure, it is also part of the pattern, can be used for the bottom contrast pattern, and can become the boundary between patterns.

Furthermore, the basic shapes composed of a single color, their overall patterns can either be the same color as the basic shape, such as NO.01ⓐⓑ Hakul (Hakku), NO.02-Backⓐ Makanji, NO.04ⓑ Msbtunux, NO.11-Backⓑⓒⓓⓔⓕⓖⓗⓘⓙⓚⓛ Klesan, NO.24-Backⓑ Malepa (as shown in Fig. 6.17), NO.27-Frontⓐⓑ

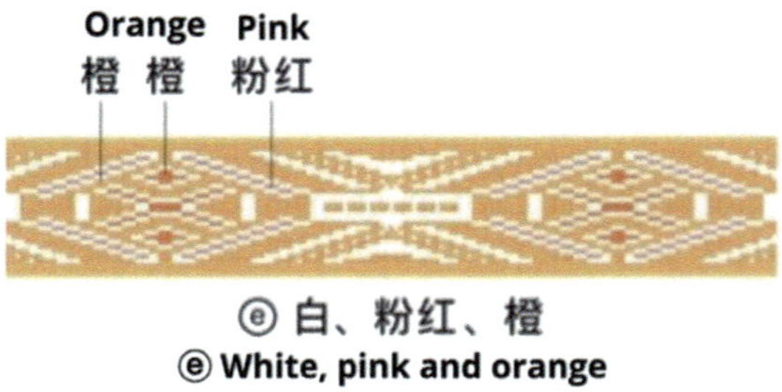

Fig. 6.15 Color performance of shaping boundaries with NO.39 as an example (drawn by this study)

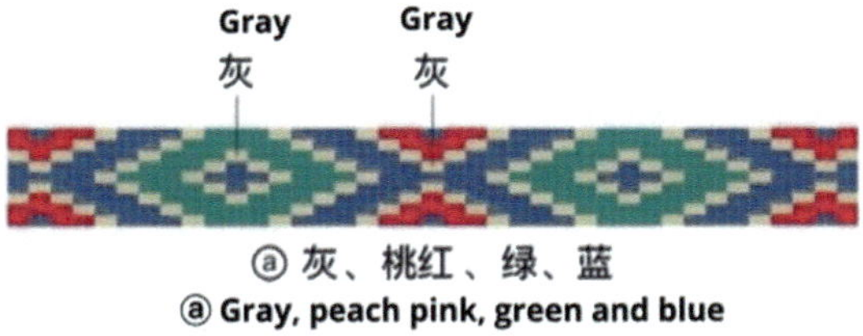

Fig. 6.16 Color performance of shaping boundaries with NO.10 as an example (drawn by this study)

Taranan, NO.30-Back ⓐⓑⓒ Plnawan, NO.33-Back ⓐⓑ Bunsui, NO.44-Front ⓑⓓⓔⓗ, NO.51-Back ⓐⓑⓒ Nanshi, NO.56.Front ⓐ Meicyubus, NO.58-Front ⓓⓕⓗⓙⓛⓜⓝⓞⓟⓠⓡⓢⓣⓦⓧⓨ Musha, NO.63-Front ⓑⓒⓓⓔ Mugu, NO.64 ⓐ Truku, etc. Alternatively, the basic shapes in a pattern can have several different color combinations, such as NO.10-Front ⓐ Klesan, NO.19 ⓐ Klesan (as shown in Fig. 6.18), NO.21 ⓐⓑ Klesan, NO.22 ⓑⓒⓓ Klesan, NO.32-Back ⓑ Bunsui, NO.35-Back ⓒ Bunsui, NO.40-Front ⓒ Bunsui, NO.47-Front ⓐⓑⓒⓓ Beishi, NO.48 ⓐⓑ Beishi, NO.49 ⓐⓑ Beishi, NO.50 ⓒⓓ Beishi, NO.53 ⓑⓓ Nanshi, NO.68 ⓑⓒ Truku, NO.69 ⓐⓑ Truku, NO.70 ⓐ Truku, NO.71 ⓑⓒ Truku, NO.73 ⓐ Truku, etc. Among them, the more unique pattern color performances are not limited to single or multiple color combinations but use color configuration to traverse and divide the pattern horizontally, forming an orderly color distribution in the same section of composite stripes, mainly concentrated in NO.21 ⓐⓑ Klesan, NO.22 ⓐⓑⓒⓓ Klesan, NO.40-Front ⓒ Bunsui, NO.48 ⓐ Beishi, NO.49 ⓑ Beishi, as shown in Figs. 6.19 and 6.20.

The color combinations of Atayal tribal textile patterns are vibrant, but the effect of these combinations can only be realized when they are linked with many complex elements. Repetition is one of the ways to enhance visual effects, presenting a regular and orderly aesthetic. This can be achieved through the continuous or alternating arrangement of the same hue, brightness, chroma, shape, area, weave pattern, and other factors. For example, in the NO.69 ⓐⓑ Truku, the diamond pattern uses only red, green, and blue colors, and the clever alternating repetition can create a rich visual effect. This study collates research samples and corresponds to the types, primary forms, compound stripe structures, and textile pattern colors, with 367 compound stripes available for comparison and analysis. Observing textile patterns and colors, it can be found that the pattern composition types of compound stripes and the primary

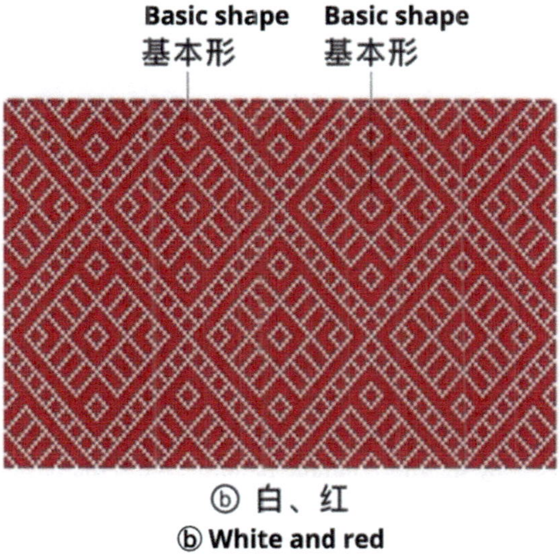

Fig. 6.17 Basic shape composed of a single color, example NO.24 (drawn by this study)

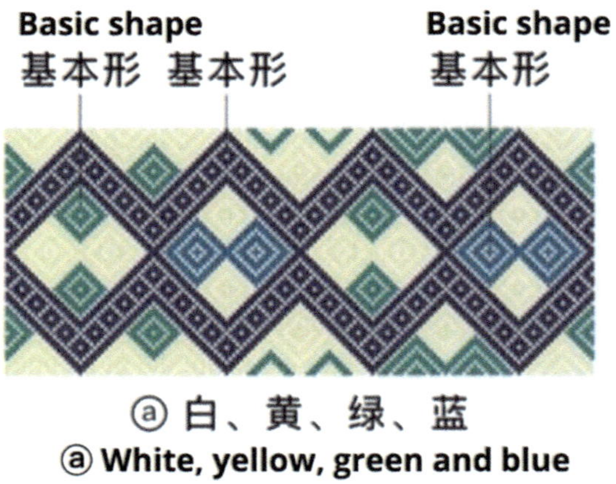

Fig. 6.18 Basic shape composed of multiple colors, example NO.19 (drawn by this study)

Fig. 6.19 Color configuration dividing the pattern, example NO.22 (drawn by this study)

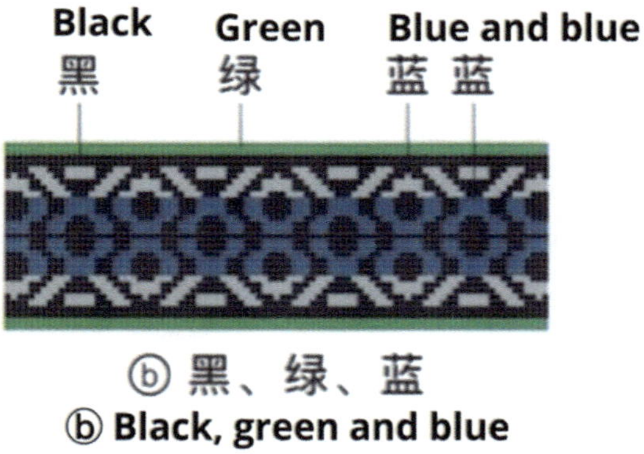

Fig. 6.20 Color configuration dividing the pattern, example NO.49 (drawn by this study)

forms of patterns have standard features, such as the NO.01ⓐⓑ Hakul (Hakku), the pattern type is a diamond pattern, and the primary form of the pattern is also a diamond. There are 136 sets of compound stripes with this same phenomenon. Among the more unique patterns, the same basic form uses two colors in a continuous or alternating repetition, such as NO.05-Frontⓕ Klesan, NO.10-Frontⓐ Klesan, and NO.14-Backⓐ Klesan (Fig. 6.21), NO.16. Frontⓒ Klesan (Fig. 6.22), NO.35-Backⓒ Bunsui (as shown in Fig. 6.23), NO.57-Frontⓐⓑ Musha (as shown in

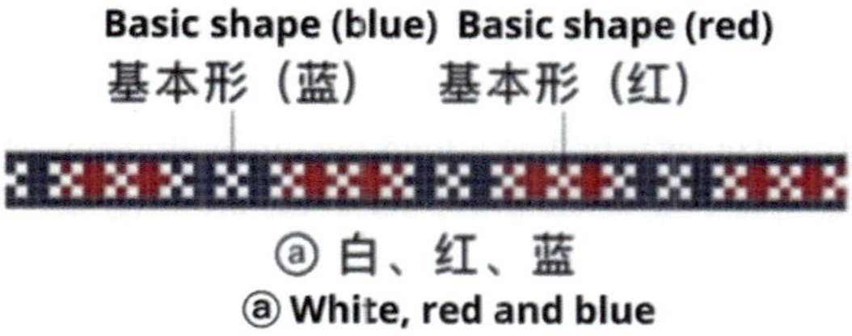

Fig. 6.21 Basic form repeated with different colors, using NO.14 as an example (drawn by this study)

Fig. 6.24), NO.68ⓐ Truku, NO.73ⓐ Truku, and other patterns, their compound stripe structures are mainly expressed in the form of horizontal separation, vertical repetition, and grouping. Although the above quantities do not account for half of the total, it can be confirmed again that the composition method of Atayal tribal textile patterns is not a fixed form, but a uniqueness created within specific rules.

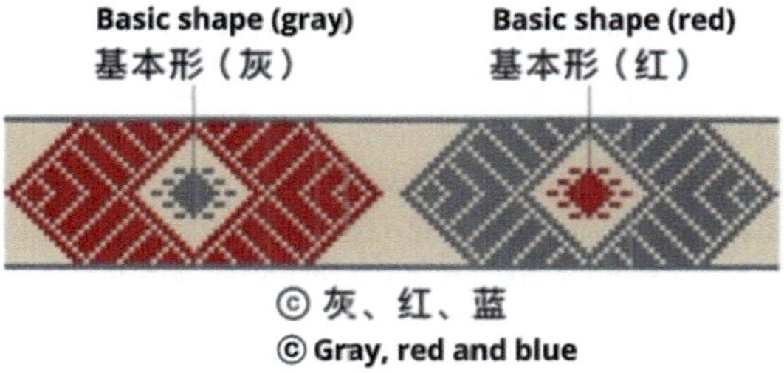

Fig. 6.22 Basic form repeated with different colors, using NO.16 as an example (drawn by this study)

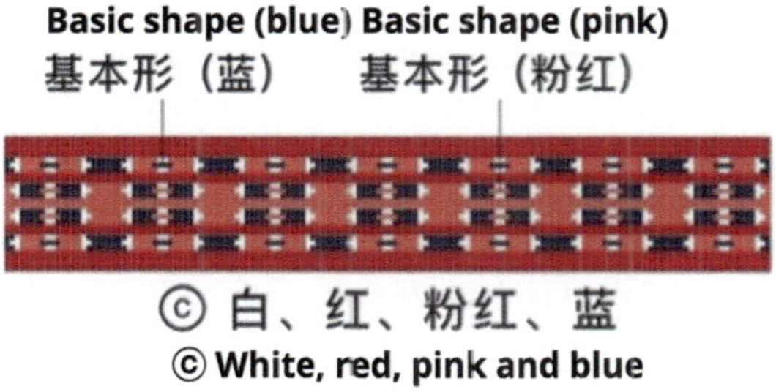

Fig. 6.23 Basic form repeated with different colors, using NO.35 as an example (drawn by this study)

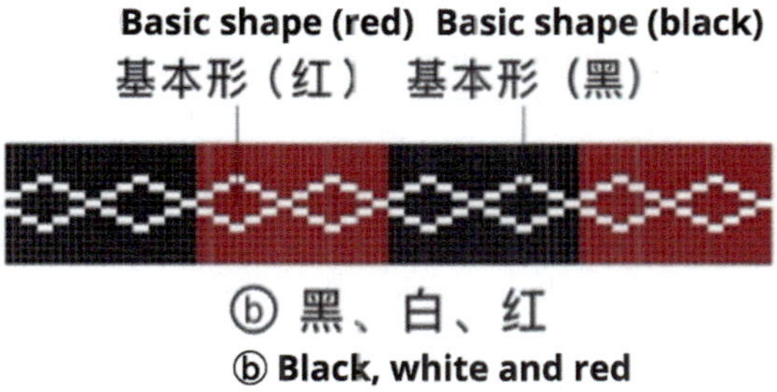

Fig. 6.24 Basic form repeated with different colors, using NO.57 as an example (drawn by this study)

6.4 Geographical Distribution of Textile Patterns

The Atayal tribe has a long history in Taiwan. The current residential areas of the various Atayal subgroups are not the original tribes of the past, and the origins narrated by the tribes within the same subset are also different. This may be due to the various geographical knowledge caused by the distance of migration over generations, coupled with the gradual estrangement from the original hometown over time and the forgetfulness of older people in the tribe, leading to the gradual blurring of geographical knowledge and many other factors. Eventually, there will be errors in oral transmission, and only a vague direction can be remembered. Nowadays, describing the complete picture through field investigation may be difficult. Therefore, this study collates the historical documents accumulated from previous related surveys as the basis, such as "The Atayal People of Klesan: Ethnographic Field Research and Study Volume I and II" [2], "Aboriginal Custom Survey Report: Volume I Atayal" [3], "Taiwan's Minority History: Atayal History" [4], "Aboriginal Survey Report: Volume VII Atayal Part II, Volume IV Seediq and Truku, Volume V Atayal Part I" [5], "Takasago Tribe Survey Book: Tribal Overview" [6], "Research on the System of Taiwan's Minority: Volume I Main Text" [7], "Taiwan's Minority History Language Culture Dictionary",[1] etc. These research works explore the geographical distribution and subgroup relationships of Atayal patterns. The scope is only for the subgroups that already have research samples, such as the Hakul (Hakku), Makanji, Msbtunux, Klesan, Malepa, and Taranan of the Tayal Sekelek; the Klesan, Plnawan, Bunsui, Beishi, Nanshi, and Meicyubus of the Tseole; the Musha, Pulibau, and Truku of the Seediq. The geographical distribution and subgroup relationships of the Atayal tribe can be imagined to be highly complex. From the records of related documents, it can be understood that many scholars at various times, through field investigation verification combined with imaginative and speculative interpretation methods, gradually linked the geographical distribution of the Atayal tribe. The separation and merger of ethnic groups caused by the migration process of each subgroup can be understood from previous research surveys. This study collates the legendary origins and migration path relationships of the Tayal and the Seediq sub-tribe, showing their complexity. Therefore, through different origins, the differences and distribution within the Atayal tribe and the migration situation of later communities can also be clarified.

Although the geographical distribution and subgroup relationships of the Atayal tribe are complex, it is not difficult to find some critical clues within them. For example, the geographical relationships of closely related subgroups influence each other. Subs at a greater distance may interact due to blood relationships or intermarriage between subgroups, resulting in some degree of exchange. Here, by using the legends of the origins and migration paths of the Atayal subgroups, we compare the subgroup textiles to explore the context of the patterns. The following focuses on the origins of the subgroups.

[1] Taiwanese Minority History, Language, and Culture Dictionary, https://reurl.cc/6Nky6O, 6th April 2023.

(1) The legendary origin of the Seqoleq subgroup is the Binsbogan Rift Rock, located in Ren'ai Township, Nantou County. The Hakul (Hakku) has a long historical origin and can be regarded as the root of the Atayal tribe. From the research samples, it was found that the styles of the three textiles, NO.01 Hakul (Hakku) (men's and women's shawls) and NO.60, NO.61 Musha (men's and women's shawls), are very similar, with NO.01ⓑ Hakul (Hakku) and NO.61ⓐ Musha having similar diamond patterns, as shown in Fig. 6.25. The Makanji originates from the Makanaqi community (Makanji) (also known as the Makanaqi) of the Hakul (Hakku). The Malepaand and Hakul (Hakku) are geographically adjacent and are said to separate from the Hakul (Hakku) to form their group. The Taranan comes from the tribe formed by the Malepa and is also one of the subgroups formed by the early Atayal migration northward. The Taranan's clothing style is similar to that of the Msbtunux, possibly due to the early Fuba crossing ridge ancient path (connecting the Atayal tribe of Fushan in Wulai Township, Taipei County with Baling in Fuxing Township, Taoyuan County), such as NO.03, NO.04 Msbtunux (men's and women's shawls) and NO.28 Taranan (women's long skirts), NO.29 Taranan (men's and women's shawls), these four textiles all have horizontal delicate patterns on the cloth.

The Klesan is not a single group in the system. It includes many outsiders, such as the Wuta community, which comes from part of the Malepa and is in Havensporu (upper reaches of the Dajia River). In the research samples, NO.18, NO.19 Klesan (women's belts), NO.20 Klesan (women's foot protection cloth), and three other textiles belong to the Wuta community. Comparing the textile patterns of the Klesan's Wuta community and the Malepa, it was found that the NO.18ⓔ Klesan and the NO.26ⓒ Malepa have similar diamond patterns, as shown in Fig. 6.26. The Jinyang community comes from the Kuru community (Kuru) (located on the right bank of the Gaogan River upstream of the Msbtunux River); Gogan is the back mountain of the Msbtunux (or called Gogan), translated as "Dgiyaq tribe" from 1895 to 1945. In addition, there are also small groups split from the Malepa, Hakul (Hakku), etc. NO.11 Klesan (men's sleeveless long clothes), NO.14 Klesan (men's sleeveless short clothes), NO.23 Klesan (men's and women's shawls) and other three textiles are from the Jinyang community. Comparing the textile patterns of the Klesan's Jinyang community and the Malepa, among them, NO.11-Backⓔ Klesan, NO.14-Backⓑ Klesan, NO.24-Backⓐ Malepa, NO.25ⓐ Malepa, etc., all have similar diamond patterns, as shown in Fig. 6.27.

Fig. 6.25 Comparison of Hakul (Hakku) and Musha patterns (drawn by this study)

Fig. 6.26 Comparison of Klesan Wuta Society and Malepa patterns (drawn by this study)

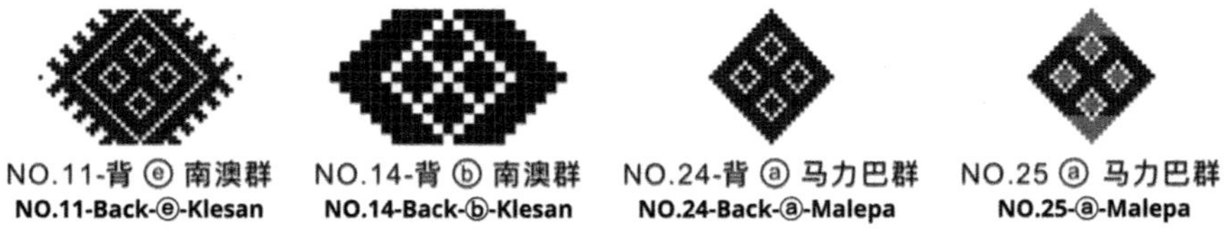

Fig. 6.27 Comparison of Klesan Jinyang Society and Malepa patterns (drawn by this study)

(2) The legendary origin of the Tseole is the Damajian Mountain, located at the junction of Hsinchu County, Miaoli County, and Taichung County. The Plnawan is a relatively small group of Atayal people, mainly composed of the Tuuda, Toroko Group (also known as Truku), Hakul (Hakku), etc., among which the Plnawan society (Perugawan) and the Hakul (Hakku) Maisidobang Society (Mastbwan) (now called Ruiyan) have close blood relations. The research samples show that NO.30 Plnawan (men's long-sleeved long clothes) and NO.01 Hakul (Hakku) (men's and women's shawls) are mainly red series clothing. Due to the small number of Plnawan textiles, it is currently impossible to compare their textile patterns with those of the Tuuda, Toroko Group, Hakul (Hakku), etc. This is only a brief description. Documents from 1895 to 1945 pointed out that the Bunsui and the Dahu Group may have intermarried with other tribes, which may be due to geographical proximity. The Pkwali and Tapiras societies, belonging to the Bunsui, also have close blood relations. According to this statement, it can indeed be found in the research samples that the styles of NO.31 Bunsui (men's sleeveless long clothes), NO.34 Bunsui (men and women's long-sleeved long clothes), and NO.35, NO.36 Bunsui (men's long-sleeved long clothes), as well as Tapiras Society's NO.32 Bunsui (men's sleeveless long clothes) and NO.37 Bunsui (men's long-sleeved short clothes) are indeed similar. Observing their textile patterns, NO.34-Front ⓐ Bunsui, NO.32- Back ⓑ Bunsui, NO.36. Front ⓒ Bunsui, and NO.37-Front ⓑ Bunsui, the above are similar patterns, as shown in Figs. 6.28 and 6.29.

The Beishi and the Nanshi are geographically close and overlapping, but the Mepasing Society, which belongs to the Nanshi, is like the Beishi culture. Later, some Mepasing Society members moved into the Beishi's Temukbonay Society. The Beishi's Sekohan Society and the Bunsui's Tapiras Society have close blood relations, and the Sekohan Society, which was initially considered part of the Beishi, half of

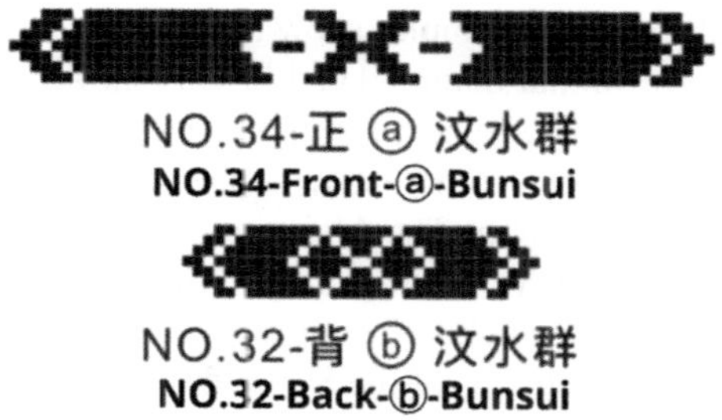

Fig. 6.28 Comparison of patterns of Bagua Li community and Dabila Si community (drawn by this study)

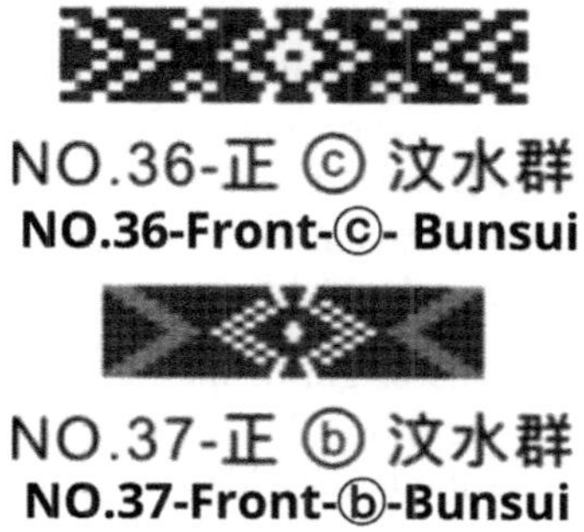

Fig. 6.29 Comparison of patterns of Bagua Li community and Dabila Si community (drawn by this study)

its members migrated from the Dahu Group. This document shows that the Bunsui, Beishi, Nanshi, and Dahu have close relationships. The observation of textile styles also shows similar styles, such as NO.39, NO.40 Bunsui (women's long-sleeved long clothes) and NO.44, NO.45 Beishi (women's long-sleeved long clothes) are classic Atayal bride's dress styles.

The Nanshi is a collection of many subgroups, which can be divided into those belonging to the Saikolik system and two or three communities belonging to the Beishi. The Qing Mountain community (Ulai-ruma) is a small group that split from the Shalamo group. Most of the Bo'ai community (Timburan) is a small group that migrated from the Masitobaon community of the Hakul (Hakku), with three households from the Meiyuan group also moving in. The Nan A Leng community (Saqauyau) is a small group that split from the Məl-qoan of the Malepa (presumably the ancestors of the Marikuo). Noteworthy descriptions in the literature from 1895 to 1945 mention that the current site of the Nan A Leng community was originally the land of the Meiyuan group. Each incoming household gave a piece of Pala aboriginal cloth (about 5 feet of hemp fabric, either white or dyed brown with betel nut) and a live pig as exchange conditions. From this, it can be understood that at that time, aboriginal cloth was used as one of the items for trade, and it is speculated that it may have been used or imitated by other subgroups. The Shao Lai Ping community (Mai-Saurai) (also known as the Shao Lai community) and the White Hair community (allegedly originally of the same tribe as the Hakul [Hakku]) migrated together. Still, it is speculated that the ancestors of the two communities may have come from the Mək-moyau of the Malepa (now known as Wangyang). If this theory is correct, the research samples show that NO.52ⓐ of the Nanshi and NO.25ⓐ of the Malepa have similar diamond patterns, as shown in Fig. 6.30.

The Meicyubus, also known as the Meicyubus community (Cyubus), formerly the Jiubusi community, comes from the Bailan community (Pskwaran) of the Jialapai group. The Meicyubus community has a close kinship relationship with the Raremewan community of the Saixia tribe. Research samples NO.55 and NO.56 of the Meicyubus (men's sleeveless long clothes) are similar in style to the traditional costumes of the Saixia tribe, with the only difference being the patterns. The Msbtunux originates from the Daba Tip Mountain, and it has been mentioned earlier that it had interactions with the Taranan due to the ancient path over the Fuba Ridge. In the research book, NO.03 and NO.04 of the Msbtunux (men's and women's shawls) are typical costumes of this subgroup. NO.04 of the Msbtunux (men's and women's shawls) records the original collection site as the Li Mao'an community (also known as the Rimogan community). Still, the Li Mao'an community should be the residence of the Taranan. This can correspond to the literature records about the interactions between the Msbtunux and Taranans due to the ancient path.

(3) The origin of the Seediq Atayal tribe's legend is the boulder of Baishishan, located within the Ren'ai Township of Nantou County. The Klesan of the Biho tribe is the oldest tribe of this branch. According to documents, it originated from the Komaipo tribe north of Musha. In the research samples, NO.07, NO.08, NO.09 Klesan (men's sleeveless long clothes), and NO.17 Klesan (women's belts) are all textiles belonging to the Biho tribe. The three textiles of NO.07, NO.08, NO.09 Klesan (men's sleeveless long clothes) all have similar mountain patterns, such as NO.07-Backⓓ Klesan, NO.08-Backⓑ Klesan, NO.09-Backⓐ Klesan. Among them, NO.09 Klesan (men's sleeveless long clothes) and NO.57 Musha (men's sleeveless long clothes) are mainly red. Another common feature is that they only have a few woven patterns. In addition, the NO.17ⓒⓘ Klesan and the NO.58-Backⓔⓕ Musha also have similar diamonds 6.31. From this, it can be inferred that the Klesan Biho tribe may be related to the Musha.

Musha suffered much damage and migrations due to patterns, as shown in Figure, the occurrence of minority anti-Japanese events from 1895 to 1945. At that time, the leader of the Mehebu tribe, Mona Rudao, led the Truwan Tgdaya group to rise, and the surviving members were forcibly moved to the Chuanzhongdao tribe (now known as Qingliu tribe) after the Musha incident. The Mehebu tribe was also split from the Truwan Tgdaya group of the Tarowan tribe, and the time of its split was approximately

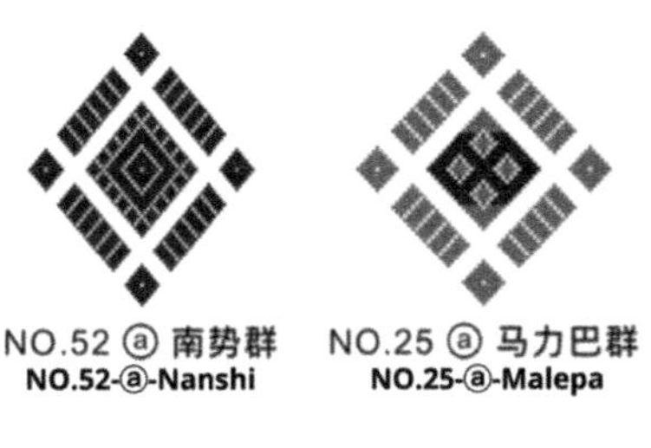

Fig. 6.30 Comparison of patterns of the Shao Lai Ping community and Malepa (drawn by this study)

after the Qacoq, Gungu, Suku, and Paran tribes. In the research samples, the NO.57 Musha (men's sleeveless long clothes) belongs to the Paran tribe; the NO.58 Musha (men's long-sleeved long clothes) belongs to the Mehebu tribe; NO.61 Musha (men's and women's shawls) belongs to the Musha; NO.62 Musha (woven mats) belongs to the Gungu tribe the patterns of these textiles vary greatly, perhaps due to the limited research samples available, making it impossible to carry out a standard comparison.

According to the literature, the early Atayal tribe migrated eastward to form different subgroups, with the Musha as the main body. Some of the eastward migrants of the Musha were the Pulibau, composed of the Truwan tribe. The current research samples only have NO.63 Pulibau (men's sleeveless long clothes), the style of this garment is quite close to the style of the North Amis group of Nanshi, especially the tassel decoration on the back of the garment, which the neighboring Nanshi Amis tribe may influence. Looking again at NO.63 Pulibau (men's sleeveless long clothes) and NO.58 Musha (men's long-sleeved long clothes), similar mountain patterns can be found, NO.63-Back ⓓ Pulibau and NO.58-Front ⓥ Musha, as shown in Fig. 6.32.

The Truku group originally lived in the Taongan area of the Musha. It later migrated to the Tarowan community (the oldest tribe of the Truku) within the Musha. Finally, those who moved east to establish along the Musha River became the Truku. According to this documentation, the Truku and the Mushas should share the same cultural background. In the research samples, it can be found that NO.64 Truku (men's breastplate) and NO.62 Musha (woven mat) have similar diamond patterns, NO.64 ⓐ Truku and NO.62 ⓐ Musha, as shown in Fig. 6.33. In addition, the Truku and Pulibaus may have similar patterns in their textiles due to their shared roots in the Taiwan community. By comparing research samples, the NO.67 Truku (women's sleeve), NO.69 Truku (women's long skirt), NO.71 Truku (men's and women's shawl), and NO.63 Pulibau (men's sleeveless long dress) indeed have similar diamond patterns, as shown in Fig. 6.34. The Truku group was originally a branch of the Atayal tribe that migrated eastward, belonging to the Seediq (formerly known as East Seediq), and has now become an independent Truku. In the west, the Musha in Ren'ai Township, Nantou County, also belongs to the Seediq (formerly known as West Seediq) and has become an independent Seediq. However, regardless, relevant literature all points out that the two have similar bloodlines, languages, and customs, and some identical textile patterns also show the connections that once existed between the branches.

Fig. 6.31 Comparison of Klesan Biho community and Musha patterns (drawn by this study)

Fig. 6.32 Comparison of Pulibau and Musha patterns (drawn by this study)

Fig. 6.33 Comparison of Truku and Musha patterns (drawn by this study)

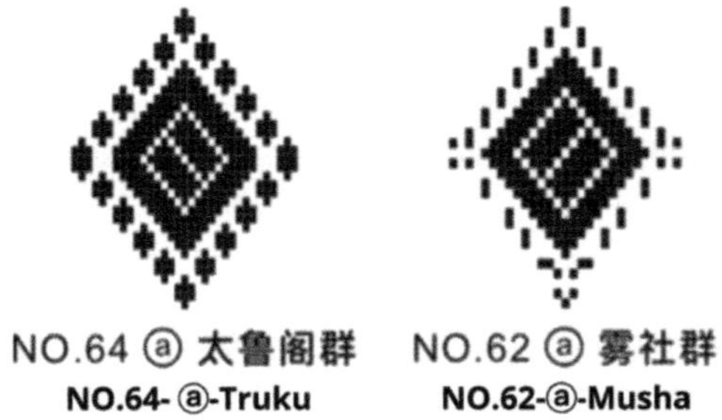

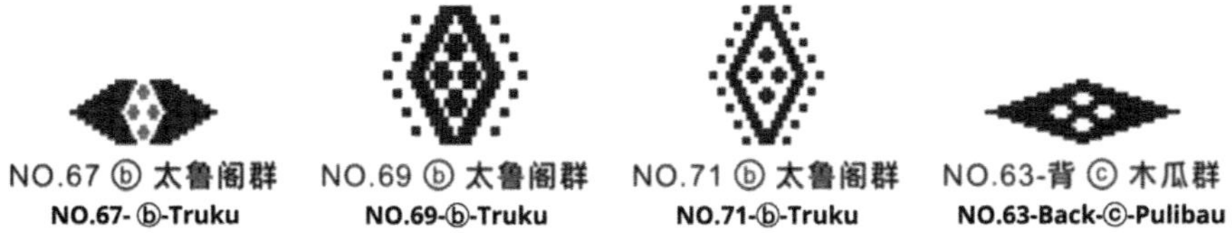

Fig. 6.34 Comparison of Truku and Pulibau patterns (drawn by this study)

This study compiles relevant literature descriptions to explore the possible geographical distribution of Atayal tribal textile patterns. This research approach is based on the hypothetical stable state presented by the Atayal subgroups in the migration process as described in the literature, that is, a comparative method based on a hypothetical standpoint. Through the viewpoint of "style standards," the patterns of two distant areas, or due to the kinship, adjacency, etc., between the subgroups, by comparing similar shapes of woven patterns, it attempts to construct a partial geographical distribution of Atayal tribal textile patterns, as shown in Fig. 6.35. Although the subgroups classified in historical literature may not necessarily conform to the cognition of the Atayal people, there may also be vague areas between each other and even the common situation of regional overlap. According to the investigation and verification of scholars at different times, and due to varying perspectives of observers, the same subgroup may also be considered another subgroup, such as "Taiwan's Minority History: Atayal History" [8], "Investigation of the Takasago Tribe: Overview of the Barbarian Society" [9]. All describe that the Tayal tribe's Nanshi was established to the east of Shaolai, a settlement composed of the Baimao and Aleng tribes, which was abolished after about 1930 and the Shaolai, Baimao, and Aleng tribes were merged into the Nanshi. "Research on the Affiliation of Taiwan's Minority Ethnic Systems: Volume One Main Text" [10]. The record believes that the Shaolai and Baimao tribes of the Nanshi, in terms of language and customs, are different from the other Nanshi belonging to the Saikolaike system, but are similar to the Beishi; therefore, they are regarded as belonging to the Beishi by their neighboring Beishi. From this, different research perspectives hold different views. The Tayal tribe itself is a tribe that has undergone migration and differentiation, and its internal sub-tribes will not objectively exist in a stable state. Therefore, traditional Tayal costumes seem to be a microcosm of the results of tribal movement and interaction with other tribes. Through the above research, it is hoped that the textile patterns can be used as a thread to explore material culture and provide another method to understand the migration process of the Tayal tribe. At the same time, it emphasizes that the clues presented by material culture are undeniably important.

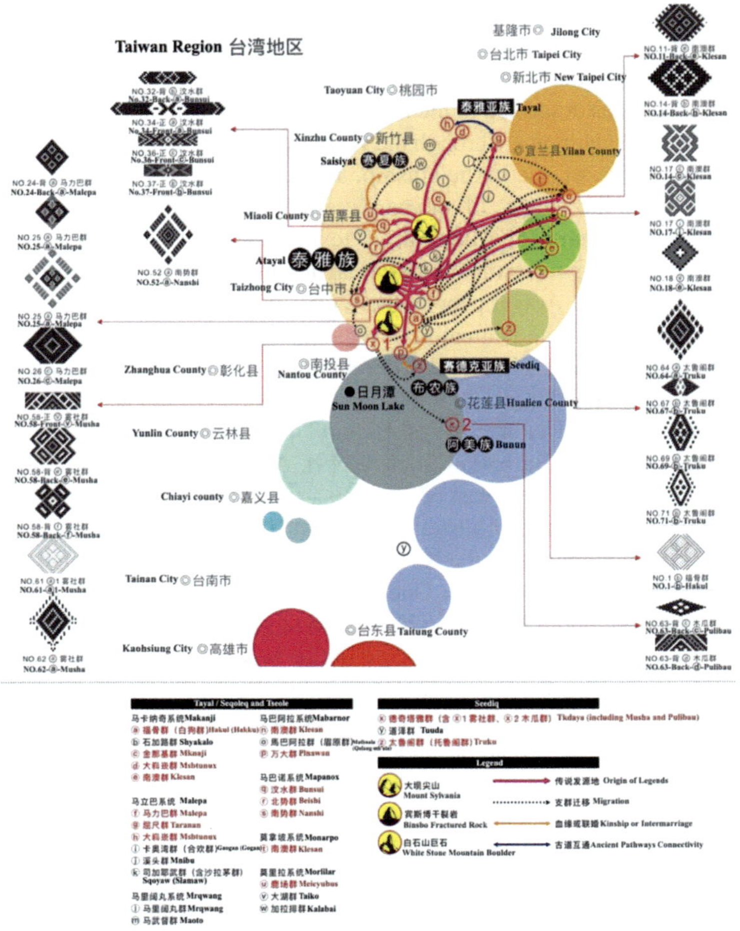

Fig. 6.35 Geographic distribution of textile patterns of the Tayal and the Seediq (drawn by this study)

References

1. Bayan Heidai: "*Exploring the Lifestyle of the Atayal: An Atayal's Personal Account*," Hsinchu County Cultural Affairs Bureau, 2002 Edition.
2. Qi-Wei Liu: "*Art Anthropology: Primitive Thinking and Creation*," Taiwan Lion Books Co., Ltd., 2002 Edition.
3. Lin Hui-xiang: "*Cultural Anthropology*," Taiwan Commercial Press, 2007 edition.
4. Ni Jian-lin: "*The Origin of Decoration: A Study of Primitive Decorative Art*," Chongqing University Press, 2007.
5. Josiane Cauquelin: "*The Aborigines of Taiwan: The Puyuma: From Headhunting to the Modern World,*" London & New York: Routledge Curzon, 2004.
6. Chen Chi-lu: "*Material Culture of the Formosan Aborigines*," Taipei: Southern Materials Center, Inc., 1998.

7. Okamura Yoshiemon, Zhang Yongxin: "*Taiwan's Aboriginal Cloth*" (Taiwan's Aboriginal Cloth), Japan Yushu Hall Co., Ltd., 1969 edition.
8. Li Sha-li: "*The Clothing Culture of the Paiwan Tribe*," Taiwan Independence Evening News Cultural Publishing Department, 1993 edition.
9. You Lan Duo You: "*Atayal Weaving Shadows*," Taiwan Rice Village Publishing House, 2004 edition.
10. Glen Dudbridge: "*Aborigines of South Taiwan in the 1880s*," Taipei: Shung Ye Museum of Minorities, Institute of Taiwan History, Academia Sinica's Taiwan, 1999.

Chapter 7
Conclusion

7.1 Research Conclusion

After comparing the research samples, analyzing the details of the textile patterns, clarifying the core patterns, and using the method of graphic interconnection to examine the overall arrangement and color of the patterns, re-examining the logical relationship between the weaving structure and color, and exploring the design thinking of Tayal textiles. By comparing the geographical distribution of sub-tribes and the corresponding relationship of textile patterns, the following discoveries are made, as follows:

From 1895 to 1945, in the study of Taiwan's minority ethnic groups, many historical documents mention that the Atayal tribe, during their migration process, likely had more contact and interaction with different ethnic groups (or other Atayal subgroups), leading to the formation of mixed-ethnicity tribes or the creation of new communities. An understanding of the relevant literature suggests that ethnic integration is one of the factors leading to cultural evolution. According to this theory, research samples are classified according to their subgroup. Most of the materials used in their textiles are homemade hemp yarn, which is then woven and tailored. In the later period, influenced by external factors, they began to use a mix of cotton, wool, and other yarns. The main weaving techniques include plain weave, twill weave, floating weave, pick weave, sandwich weave, braiding, etc. The patterns of the textiles can be said to be a cultural representation of the minority ethnic groups. During the production process, through the creative consciousness or imagination of the weaver, they convey the symbolic nature of the whole society. The visually apparent pattern shapes play a special significance. Comparing the weaving and dyeing activities of the Atayal tribe, they are all led by women, who develop and pass down the material, showing the characteristics of the ethnic group. Looking at the samples, the patterns and colors of Atayal textiles present a diverse representation, with similar styles in the same subgroup's clothing, while also retaining some features of primitive art, using simple or complex woven patterns to reflect and

P. H. Wang and J. Li, *Interpreting the Evolution of Atayal Tribal Textile Patterns*,
https://doi.org/10.1007/978-981-96-5451-2_7

describe cultural objects or symbols, such as the diamond shape commonly found in tribes symbolizing ancestral spirits' eyes. Perhaps the tribes thinking is transformed into specific points, lines, surfaces, and other abstract geometric shapes, and due to the limitations of weaving tools and techniques, similar woven pattern structures are presented between subgroups, forming commonalities. Atayal textile patterns, from the samples, can be summarized into 12 types: changing patterns, diamond patterns, circular patterns, mortar-shaped patterns, zigzag patterns, square patterns, mountain-shaped patterns, cross patterns, X-shaped patterns, net-like stripes, dot-like stripes, and stripes.

The more common shapes include diamond patterns, stripes, mountain-shaped patterns, cross patterns, zigzag patterns, X-shaped patterns, circular patterns, etc. The textiles with a more significant number of patterns are mainly men's clothing, such as five patterns in NO.06, NO.31 men's sleeveless long clothes; NO.13, NO.14 men's sleeveless short clothes; NO.30, NO.35 men's long-sleeved long clothes, six patterns in NO.36, NO.58 men's long-sleeved long clothes, eight patterns in NO.5 men's sleeveless long clothes, mainly appearing in Klesan, Plnawan, Bunsui, and Mushas. The above patterns are distributed among various subgroups. Still, different subgroups have found the same shapes, such as the NO.14-Backⓑ Klesan (men's sleeveless short clothes) and the NO.51-Backⓑ Nanshi (men's long-sleeved long garments). The Klesan and Nanshi are far apart and have no kinship or marriage ties. Still, the same diamond pattern appears, which is speculated to be due to the limitations of weaving tools and techniques, resulting in similar woven pattern structures. Weaving technology is a crucial factor influencing textile patterns; this principle generally applies to many minority ethnic textiles and has the same characteristics as Taiwan's minority ethnic groups, such as diamonds, octagonal flowers, and other patterns, which are pervasive woven patterns.

The design patterns of general textiles often include basic shapes such as diamonds, circles, rectangles, mountain shapes, cross shapes, X shapes, diagonal line shapes, etc., among which diamonds, mountain shapes, and cross shapes are used more frequently. Why do these patterns often appear? It is speculated that geometric patterns should be technical weaving features, such as " + " weaving, the unit structure most likely to occur after the warp and weft intersect in textiles. Subsequently, with this shape as the center of the weaving composition, it can gradually extend to form various basic shapes or other more complex geometric patterns, such as NO.01ⓑ Hakul (Hakku) (men and women's shawls), NO.04ⓐ Msbtunux (men and women's shawls) and other weavings. The pattern's basic shape is the weaving's core, readily revealed by color matching. The basic shape is a standard essential element of many complex patterns, and the same basic shape can be observed in the patterns of different subgroups. This basic shape  can be seen in the same patterns in the Makanji, Msbtunux, Klesan, Malepa, Bunsui, Beishi, Nanshi, Musha, Pulibau, Truku, etc. From this, it can be inferred that the more the same basic shapes, the closer the relationship between the subgroups seems to be. This implies that many Atayal subgroups have the exact origin and may have dispersed in the northern mountains of Taiwan due to migration, creating new patterns in the new environment but still retaining some patterns, which can be used to explore the relationship between subgroups.

In addition, from the pattern samples of Atayal textiles, it is found that the weaving patterns are like the rules of formalized patterns. Especially in textiles such as NO.22 Klesan (men and women's shawls), NO.61 Musha (men and women's shawls), and NO.70 Truku (men and women's shawls), this phenomenon is present. The shaping features of the textile patterns mentioned above are mainly based on the internal structure of the diamond pattern, and the weaving process is done through modular replacement, constructing many different shapes. It is inferred that this phenomenon is not accidental but planned before weaving. The reason is that the "Survey Report on Aboriginal Customs: Volume One Atayal" [1] mentions that each tribe has established ownership norms for weaving patterns, and the changes in patterns tend to be more modular to break through the patterns and develop patterns different from others, thus creating many new styles of weaving. Comparing research samples, the NO.60 Musha (men and women's shawls) has fewer pattern styles than the NO.61 Musha (men and women's shawls) and NO.71 Truku (men and women's shawls), NO.73 Truku (men and women's shawls) also have fewer pattern styles than NO.70 Truku (men and women's shawls). From this, the same piece of textile has more patterns than other textiles of the same style, which should be used for purposes like education, inheritance, or showing off technology. It is constructed as a textile, like a pattern database.

The textile patterns are arranged horizontally or vertically to form composite stripes. The formation of composite stripes includes single, vertical separation, horizontal separation, vertical repetition, horizontal repetition, vertical intersection, horizontal intersection, vertical connection, horizontal connection, grouping, and X-shaped arrangement. Observing the arrangement of composite stripes, one can see no fixed concentration in any one group. The statistics of the number of research samples show that the horizontal separation structure is widely distributed in the textiles of all groups. The horizontal repetition structure is second, distributed in the textiles of groups such as NO.01ⓑ Hakul (Hakku), NO.03ⓐ Msbtunux, NO.10-Backⓐⓗⓘ Klesan, NO.30-Rightⓑ Plnawan, NO.31-Backⓔⓕ Bunsui, NO.44-Backⓓ Beishi, NO.54ⓐ Nanshi, NO.58-Rightⓖⓨ Musha, NO.63-Rightⓐ Pulibauya, NO.67ⓑ Truku, etc. The composite stripes of the Atayal tribe's textiles are composed of many similar basic shapes. The weaving process expands outward continuously and has a high degree of change after combining into patterns. The patterns consisting of a single or more than two basic shapes form more layered pattern changes or more complex pattern structures, such as NO.39 Bunsui (women's long-sleeved long dress), NO.42 Beishi (men's long-sleeved long dress), NO.46 Beishi (women's long-sleeved short dress) and other group textiles, can be described as the most complex composite stripe structure.

Analyzing the composite stripes of the Atayal tribe's textiles, we find that the repetition of unit patterns indeed exists in the arrangement rules. The arrangement of composite stripes can be divided into reflection, such as staggered reflection, interval reflection, and leapfrog reflection. Repetition, such as translation repetition, interval repetition, leapfrog repetition, vertical repetition, and group repetition. Due to the development of the repetition of unit patterns, limited by the weaving area or the truncation of patterns, it confuses defining the arrangement rules, such as NO.01ⓐⓑ Hakul (Hakku) (men's and women's shawls), NO.22ⓑⓒⓓ Klesan (men's and women's shawls), NO.61ⓐⓑⓒⓓⓔⓕⓖⓗ Musha (men's and women's shawls), NO.56-Backⓐⓑ Meicyubus (men's sleeveless long dress), NO.62ⓐⓑⓒ Musha (weaving mat), NO.70ⓐ Truku (men's and women's shawls) and other composite stripe structures. However, the above arrangement rules are in place. Still, because the same section of composite stripes appears with several different patterns staggered, violating the principle of the same shape to form composite stripes, this phenomenon is also one of the characteristics of the Atayal tribe's textiles. The textile patterns are orderly and rich in texture, yet full of changes, which is the flat composition of clothing design. The composite stripes of the Atayal tribe's textiles, although generally arranged symmetrically, only have some irregular weaving patterns. Whether intentional or unintentional, on the one hand, it is also due to the manual weaving process, which cannot fully consider the accuracy of the weaving pattern, thus forming some differences or mistakes in the pattern, but ultimately cannot erase the unique ethnic style of the Atayal tribe's textiles.

The color combinations of compound stripes, colorless black, grey, white, and colored red, pink, orange, yellow, green, blue, purple, brown, etc., are all commonly used in Atayal textiles. The statistical analysis of the research samples shows that the most frequently used colors in compound stripes are black, grey, white, red, and blue. The color combinations of compound stripes can be divided into two, three, four, five, six, and seven color combinations. Colorless black, grey, and white can be said to be the main colors used in compound stripes, but in the four, five, six, and seven color combinations, other colors become the main colors and outnumber the colorless ones. The seven-color combination of compound stripes only appears in the textiles of the Bunsui and Beishis. Furthermore, black, grey, white, or brown should be the earliest color range used in Atayal textiles; environmental resources mainly limit these colors. The collection time recorded by the current research samples is approximately between 1895 and 1945, and there is no way to verify the Atayal textiles before this time, so we can only make assumptions here. In addition, before entering the museum to collect artifacts, field investigations were also conducted, interviewing Atayal people who were engaged in weaving skills in areas such as Ren'ai Township, Jianshi Township, Taian Township, and Klesan Township, asking about how to obtain dyeing materials. Still, for colors other than black, grey, white, and brown, most of them mentioned exchange or modernly obtained chemical dyes; perhaps other color dyes have been reduced in use or may have been forgotten due to inheritance issues. The colors of Atayal textiles are often considered to be mainly black, white, and red. Observing the color matching of textile patterns can distinguish the group to which they belong, such as NO.04ⓑ Msbtunux (men's and women's shawls) mainly in black and white; NO.22ⓐ Klesan (men's and women's shawls) has black, peach red, yellow, blue and other color systems, the two have obvious differences in textile color, and it is this difference that constitutes the characteristics of Atayal subgroup textiles. Each subgroup does have different degrees of preference for colors, just in terms of black, white, and red, there are different brightness and chroma, among which white is the color presented after ramie bleaching; some are not completely white but retain part of the original color of ramie, or even use the original color of ramie as the base color, this part can be understood from the research samples. The color matching of the compound stripe structure of Atayal textiles, the color performance of the pattern shaping boundary has no definite standard, mainly following a dark and light color arrangement combination, among which only a small part appears similar color matching performance, such as the pattern of NO.39- Zhengⓔ Bunsui (women's long-sleeved long dress). The simple pattern shaping boundary generally uses black, grey, white, and other color systems as the pattern separation; these isolation colors are sometimes also the pattern base color (base cloth color). The color matching of textile patterns often adopts repeated color matching, such as the basic shape of diamond patterns, which is also a diamond. Its basic shape uses two different colors in a continuous or alternating way to repeat, thereby creating a multi-layered pattern effect. Different types of textile patterns have different color-matching methods, but there is no fixed pattern in the use of color combinations. The patterns in the textiles of the Atayal tribe can verify the interactive relationships generated by the migration process of some subgroups. For example, the Wuta community of the Klesan, which originates from the legend of the Binspo Cracked Rock, comes from part of the Malepa. Comparing the textile patterns of the two, it is found that the NO.18ⓔ Klesan (women's belt) and NO.26ⓒ Malepa (tobacco bag) have similar diamond patterns. The Bagua Li community and the Dabulas community, originating from the legend of the Dam Tip Mountain, have close relations due to kinship. The styles of their textiles, such as NO.31 Bunsui (men's sleeveless long robe), NO.34 Bunsui (men's and women's long-sleeved long robe), and NO.35, NO.36 Bunsui (men's long-sleeved long robe), as well as NO.32 Bunsui (men's sleeveless long robe) and NO.37 Bunsui (men's long-sleeved short robe) from the Dabulas community, are similar. For example, the Wu community, which originates from the legend of the White Stone Mountain megalith, is the main source of the early eastward migration of the Atayal tribe. The Truku, which migrated eastward from the Wu community's Taiwan community, has a common cultural background. From the research samples, it can be found that NO.64 Truku (men's chest cover) and NO.62 Wu community (woven mat) have similar diamond patterns. The unity or change of textile patterns is more prominent in the inheritance and development of traditional Atayal costumes. Also, to maintain the overall image of the subgroup costumes and avoid similar patterns, they must constantly seek changes. Generally, the internal organization of pattern formation can be seen as a combination of logical structure and ethnic emotions. The evolution of Atayal textiles partly changes with time and space. In the process of cultural exchange, not only are there changes in patterns, but also the addition of foreign weaving materials and dyeing materials, as mentioned earlier. These factors will lead to innovation in traditional costumes. Still, the traditional Atayal costumes have not changed much in form, mainly showing the phenomenon of pattern flow caused by subgroup interaction. In terms of pattern form, they still retain very abstract geometric patterns, but it isn't easy to interpret their meaning. These patterns generally have continuous and repetitive expressions, and some patterns are only used in specific positions, such as the pattern of NO.39-Frontⓑ Bunsui (women's long-sleeved long robe), which only appears in the sleeve part. Overall, the patterns of Atayal textiles seem to have a master–slave relationship. The features of the main patterns are usually larger and more eye-catching, or the color performance is more prominent, such as the NO.36-Backⓐ Bunsui (men's long-sleeved long robe), and not all patterns can be placed arbitrarily. The style of traditional Atayal costumes is to gradually form styles, patterns, and colors with national characteristics and regional features within a certain area, which is also a specific manifestation of the nationalization, subgroupisation, and regionalization of costumes.

7.2 Research Reflection

This research began with fieldwork when I was fortunate enough to participate in a joint investigation activity in Taiwan by the Kobe University of Art and Design in Japan (Comparative Study of the Structure of Traditional Textile Patterns and Colors of Taiwanese Minority Tribes). Initially, I hoped that through the process of fieldwork, I could extensively collect textiles from Taiwanese minorities as primary data for research samples. However, after three years of investigation during winter and summer vacations, I found that most of the minority textiles had already been collected by museums, and some were even in overseas museums. Although traditional Paiwan and Rukai textiles can still be collected in some tribes in Kaohsiung County and Pingtung County, the number and style of samples available for research needs are still slightly insufficient. Therefore, I eventually returned to museum research, using current museum collections as the scope of research. Despite the previous fieldwork, although I could not obtain suitable research samples, I at least have a general understanding of the current situation of Taiwanese minorities and contemporary minority clothing.

Realizing that fieldwork could not collect samples needed for research, I turned to museums as the scope of the investigation. First, I started with the Taichung County Cultural Centre. I then successively obtained research samples from the Showa Museum of Ancient Art, Tenri University's Tenri Reference Museum, Shunyi Taiwanese Minority Museum, National Taiwan University Anthropology Museum, Taiwan Museum, and Taiwan Prehistoric Culture Museum. Currently, the museum collections related to Taiwanese minority textiles, including the groups, types, and quantities, are not in the minority, and it is impossible to study all Taiwanese minority groups comprehensively in a short time. Considering the planning of the research schedule, at this stage, I choose to start with the Atayal tribe as the starting point for the study of minority textiles and gradually involve the study of traditional costumes of other groups in the future.

Shifting from fieldwork to museum collections as the subject of investigation was only sometimes smooth, and of course, there were cases of rejection. The reasons for refusal were that some units believed that the textiles of Taiwanese minorities or the Atayal tribe had been studied by scholars and experts before, and many research results had been accumulated. Whether they need to be studied again, such doubts still exist. However, suppose the unit accepts the research investigation. In that case, it usually applies the research plan and explains the difference between its research content and previous research, and usually gets a response to accept the investigation. This shows that research on the material culture of Taiwanese minorities is still worth looking forward to. This study believes that the more difficult part, and the need for time comparison, is mainly the background information on the cultural relics. Museums with historical significance or scale have complete records of their collections' basic information. Only a tiny part of the textiles must be cross-compared to confirm their sub-group further. Even some textiles, due to the location of the collection at the time, were mistakenly believed to belong to the traditional costumes of that tribe, which required more time and multiple verifications to confirm their affiliation.

Research analysis methods have also been learned through multiple attempts and referenced in joint investigations with the Japanese side. Moreover, combining the knowledge of visual communication design with the perspective of anthropology has been recognized in the research submissions at this stage. Of course, the background knowledge of Taiwan's minority ethnic textiles, including the part explaining weaving techniques, still relies on the guidance and assistance provided by textile experts to present the basic information of the textiles accurately. Therefore, research topics like this must combine multiple professional fields to avoid omissions.

Reference

1. Taiwan Old Custom Survey Association: "*Survey Report on Aboriginal Customs: Volume One Atayal Tribe (番族惯习调查报告书:第一卷泰雅族),*" translated by Taiwan Academia Sinica Institute of Ethnology, Taiwan Academia Sinica Institute of Ethnology, 1996 edition.

Appendix 1
Types of Atayal Weaving (Sorted by Subgroups)

Object name/group affiliation, geographical location	Time and place of museum collection (acquisition)	Sample source
NO.01 Shawls for men and women Hakul (Hakku) (Hakul (Hakku)): Renai Township, Nantou County	June 1978, Upper Beigang River in the mountains of northern Taiwan (Faxiang Village, Ren'ai Township, Nantou County)	Tenri Reference Center Affiliated to Tenri University
NO.02 Men's sleeveless long coat Makanji: Jianshi Township and Wufeng Township, Hsinchu County; Nanzhuang Township, Miaoli County	1895–1945 (collected), Xiuluan Village, Jianshi Township, Hsinchu County (Xiuluan, Kongxi)	Taiwan Museum
NO.03 Shawls for men and women Dalikan Group: Fuxing Township, Taoyuan County	March 1966, Northern Taiwan Mountains	Tenri Reference Center Affiliated to Tenri University
NO.04 Shawl for men and women大The Rock Group: Fuxing Township, Taoyuan County	1895–1945, Wulai Township, Taipei County	Taiwan Museum
NO.05 Men's sleeveless long coat Klesan: Klesan Township and Datong Township in Yilan County	February 1969, Taipei City	Tenri Reference Center Affiliated to Tenri University
NO.06 Men's sleeveless long coat Klesan: Klesan Township and Datong Township in Yilan County	July 1930 (collected), Meiyuan Village, Tai'an Township, Miaoli County	National Taiwan University Museum of Anthropology

(continued)

P. H. Wang and J. Li, *Interpreting the Evolution of Atayal Tribal Textile Patterns*,
https://doi.org/10.1007/978-981-96-5451-2

(continued)

Object name/group affiliation, geographical location	Time and place of museum collection (acquisition)	Sample source
NO.07 Men's sleeveless long coat Klesan: Klesan Township and Datong Township in Yilan County	March 1936 (collected), Bihou Village, Klesan Township, Yilan County	National Taiwan University Museum of Anthropology
NO.08 Men's sleeveless long coat Klesan: Klesan Township and Datong Township in Yilan County	March 1936 (collected), Bihou Village, Klesan Township, Yilan County	National Taiwan University Museum of Anthropology
NO.09 Men's sleeveless long coat Klesan: Klesan Township and Datong Township in Yilan County	1895–1945 (collected), collection location unknown (like the style of Bihou Village, Klesan Township, Yilan County)	Taiwan Museum
NO.10 Men's sleeveless long coat Klesan: Klesan Township and Datong Township in Yilan County	January 1970, Northern Taiwan Mountains	Tenri Reference Center Affiliated to Tenri University
NO.11 Men's sleeveless long coat Klesan: Klesan Township and Datong Township in Yilan County	March 1936 (collected), Jinyang Village, Klesan Township, Yilan County	National Taiwan University Museum of Anthropology
NO.12 Men's sleeveless long coat Klesan: Klesan Township and Datong Township, Yilan County	1895–1945 (collected), collection location unknown (like the style of Jinyang Village, Klesan Township, Yilan County)	Taiwan Museum
NO.13 Men's sleeveless short coat Klesan: Klesan Township and Datong Township, Yilan County	June 1998 (collected) (weaving date 1897), Xiulin Township, Hualien County	Taichung County Cultural Center
NO.14 Men's sleeveless short coat Klesan: Klesan Township and Datong Township, Yilan County	March 1936 (collected), Jinyang Village, Klesan Township, Yilan County	National Taiwan University Museum of Anthropology
NO.15 Men's and women's long-sleeved shorts Klesan: Klesan Township and Datong Township, Yilan County	The collection time is unknown (about 1895–1945), the collection location is unknown	Showa-kan Ancient Art (Private Collection)

(continued)

(continued)

Object name/group affiliation, geographical location	Time and place of museum collection (acquisition)	Sample source
NO.16 Men's long-sleeved shorts Klesan: Klesan Township and Datong Township, Yilan County	1895–1945 (collected), collection location unknown	Taiwan Museum
NO.17 Women's belt Klesan: Klesan Township and Datong Township in Yilan County	October 1952 (collected), Bihou Village, Klesan Township, Yilan County	National Taiwan University Museum of Anthropology
NO.18 Women's belt Klesan: Klesan Township and Datong Township in Yilan County	October 1952 (collected), Bihou Village, Klesan Township, Yilan County	National Taiwan University Museum of Anthropology
NO.19 Women's belt Klesan: Klesan Township and Datong Township in Yilan County	October 1952 (collected), Bihou Village, Klesan Township, Yilan County	National Taiwan University Museum of Anthropology
NO.20 Women's foot protection cloth Klesan: Klesan Township and Datong Township in Yilan County	October 1952 (collected), Bihou Village, Klesan Township, Yilan County	National Taiwan University Museum of Anthropology
NO.21 Shawls for men and women Klesan: Klesan Township and Datong Township in Yilan County	June 1998 (collected) (weaving period 1937), Heping Village, Xiulin Township, Hualien County	Taichung County Cultural Center
NO.22 Shawls for men and women Klesan: Klesan Township and Datong Township in Yilan County	June 1998 (collected), Klesan Township, Yilan County	Taichung County Cultural Center
NO.23 Shawls for men and women Klesan: Klesan Township and Datong Township in Yilan County	March 1936 (collected), Jinyang Village, Klesan Township, Yilan County	National Taiwan University Museum of Anthropology
NO.24 Men's sleeveless long coat Malepa: Ren'ai Township, Nantou County	March 1929 (collected), collection location unknown (like the style of Lixing Village, Ren'ai Township, Nantou County)	National Taiwan University Museum of Anthropology

(continued)

(continued)

Object name/group affiliation, geographical location	Time and place of museum collection (acquisition)	Sample source
NO.25 Men's chest pocket Malepa: Ren'ai Township, Nantou County	1895–1945 (collected), collection location unknown (like the style of Lixing Village, Ren'ai Township, Nantou County)	Taiwan Museum
NO.26 Tobacco Pouch Malepa: Ren'ai Township, Nantou County	1895–1945 (collected), collection location unknown (like the style of Lixing Village, Ren'ai Township, Nantou County)	Taiwan Museum
NO.27 Men's sleeveless long coat Taranan: Wulai Township, Taipei County	1895–1945 (collected), collection location unknown	Taiwan Museum
NO.28 Women's long skirt Taranan: Wulai Township, Taipei County	June 1993 (collected) (weaving date 1920), collection location unknown	Taichung County Cultural Center
NO.29 Shawl for men and women Taranan: Wulai Township, Taipei County	2006 (collected), Nanshi District, Taichung County	Taiwan Museum of Prehistory
NO.30 Men's long-sleeved long coat Wan Daqun: Renai Township, Nantou County	November 1929 (collected), Plnawan Tribe, Qinai Village, Renai Township, Nantou County	National Taiwan University Museum of Anthropology
NO.31 Men's sleeveless long coat Bunsui: Taian Township, Miaoli County	July 1930 (collected), Baguali Tribe, Bagua Village, Tai'an Township, Miaoli County	National Taiwan University Museum of Anthropology
NO.32 Men's sleeveless long coat Bunsui: Taian Township, Miaoli County	1945 (Collected), Yuandun Tribe, Jinshui Village, Tai'an Township, Miaoli County	Taiwan Museum
NO.33 Men's sleeveless long coat Bunsui: Taian Township, Miaoli County	1895–1945 (collected), Yuandun Tribe, Jinshui Village, Tai'an Township, Miaoli County	Taiwan Museum
NO.34 Men's and women's long-sleeved long clothes Bunsui: Taian Township, Miaoli County	Unknown collection date (approximately 1895–1945), northern Taiwan mountains	Tenri Reference Center Affiliated to Tenri University
NO.35 Men's and women's long-sleeved long clothes Bunsui: Taian Township, Miaoli County	July 1930 (collected), Baguali Tribe, Bagua Village, Tai'an Township, Miaoli County	National Taiwan University Museum of Anthropology

(continued)

(continued)

Object name/group affiliation, geographical location	Time and place of museum collection (acquisition)	Sample source
NO.36 Men's long-sleeved long coat Bunsui: Taian Township, Miaoli County	July 1930 (collected), Baguali Tribe, Bagua Village, Tai'an Township, Miaoli County	National Taiwan University Museum of Anthropology
NO.37 Men's long-sleeved shorts Bunsui: Taian Township, Miaoli County	1895–1945 (collected), Yuandun Tribe, Jinshui Village, Tai'an Township, Miaoli County	Taiwan Museum
NO.38 Men's chest pocket Bunsui: Taian Township, Miaoli County	1993 (collected), Heping Village, Xiulin Township, Hualien County (like the style of Jinshui Village, Taian Township, Miaoli County)	Taiwan Museum of Prehistory
NO.39 Women's long-sleeved dress (bridal dress) Bunsui: Taian Township, Miaoli County	February 1979, Yuandun Tribe, Jinshui Village, Tai'an Township, Miaoli County	Tenri Reference Center Affiliated to Tenri University
NO.40 Women's long-sleeved dress (bridal dress) Bunsui: Taian Township, Miaoli County	February 1979, the upper reaches of Da'an River in Miaoli County	Tenri Reference Center Affiliated to Tenri University
NO.41 Men's long-sleeved coat Beishi: Taian Township, Miaoli County, Heping Township, Taichung County	June 1998 (collected), Tianlun Village, Heping Township, Taichung County	Taichung County Cultural Center
NO.42 Men's long-sleeved coat Beishi: Taian Township, Miaoli County, Heping Township, Taichung County	January 1993 (collected), Tai'an Township, Miaoli County	Taiwan Museum of Prehistory
NO.43 Men's open back pants Beishi: Taian Township, Miaoli County, Heping Township, Taichung County	February 1979, the upper reaches of Da'an River in Miaoli County	Tenri Reference Center Affiliated to Tenri University
NO.44 Women's long-sleeved dress (bridal dress) Beishi: Taian Township, Miaoli County, Heping Township, Taichung County	The collection time is unknown (about 1895–1945), the collection location is unknown (like the style of Meiyuan Village, Tai'an Township, Miaoli County)	Taichung County Cultural Center

(continued)

(continued)

Object name/group affiliation, geographical location	Time and place of museum collection (acquisition)	Sample source
NO.45 Women's long-sleeved dress (bridal dress) Beishi: Taian Township, Miaoli County, Heping Township, Taichung County	June 1998 (collected), Tiangou Tribe, Meiyuan Village, Tai'an Township, Miaoli County	Taichung County Cultural Center
NO.46 Women's long-sleeved shorts Beishi: Taian Township, Miaoli County, Heping Township, Taichung County	June 1978, Miaoli County in the mountainous area of northern Taiwan	Tenri Reference Center Affiliated to Tenri University
NO.47 Women's long-sleeved shorts Beishi: Taian Township, Miaoli County, Heping Township, Taichung County	The collection time is unknown (about 1895–1945), the collection location is unknown (like the style of Meiyuan Village, Tai'an Township, Miaoli County)	Shunyi Museum of Taiwan's Ethnic Minorities
NO.48 Women's foot protection cloth Beishi: Taian Township, Miaoli County, Heping Township, Taichung County	June 1993 (collected), Daguan Village, Heping Township, Taichung County	Taichung County Cultural Center
NO.49 Women's foot protection cloth Beishi: Taian Township, Miaoli County, Heping Township, Taichung County	February 1979, Miaoli County in the mountainous area of northern Taiwan	Tenri Reference Center Affiliated to Tenri University
NO.50 Shawls for men and women Beishi: Taian Township, Miaoli County, Heping Township, Taichung County	June 1993 (collected), Tianlun Village, Heping Township, Taichung County	Taichung County Cultural Center
NO.51 Men's long-sleeved coat Nanshi: Heping Township, Taichung County	1895–1945 (collected), Nanshi Village, Heping Township, Taichung County	Taiwan Museum
NO.52 Men's chest pocket Nanshi: Heping Township, Taichung County	1895–1945 (collected), Heping Township, Taichung County	Taiwan Museum
NO.53 Women's long skirt Nanshi: Heping Township, Taichung County	June 1998 (collected) (weaving year 1927), Daguan Village, Heping Township, Taichung County	Taichung County Cultural Center

(continued)

(continued)

Object name/group affiliation, geographical location	Time and place of museum collection (acquisition)	Sample source
NO.54 Women's foot protection cloth Nanshi: Heping Township, Taichung County	The collection time is unknown (about 1895–1945), the collection location is unknown (like the style of Nanshi Village, Heping Township, Taichung County)	Showa-kan Ancient Art (Private Collection)
NO.55 Men's sleeveless long coat Deer farms: Nanzhuang Township and Taian Township, Miaoli County	The collection time is unknown (about 1895–1945), the collection location is unknown (like the style of the old Fengmei tribe in Donghe Village, Nanzhuang Township, Miaoli County)	Showa-kan Ancient Art (Private Collection)
NO.56 Men's sleeveless long coat Deer farms: Nanzhuang Township and Taian Township, Miaoli County	July 1930 (collected), Old Fengmei Tribe, Donghe Village, Nanzhuang Township, Miaoli County	National Taiwan University Museum of Anthropology
NO.57 Men's sleeveless long coat Musha: Renai Township, Nantou County	1895–1945 (collected), Datong Village, Ren'ai Township, Nantou County	Taiwan Museum
NO.58 Men's long-sleeved long coat Musha: Renai Township, Nantou County	1929 (collected), Lushan, Yingying Village, Renai Township, Nantou County	National Taiwan University Museum of Anthropology
NO.59 Women's long-sleeved shorts Musha: Renai Township, Nantou County	June 1998 (collected), Tongmen Village, Xiulin Township, Hualien County	Taichung County Cultural Center
NO.60 Shawls for men and women Musha: Renai Township, Nantou County	June 1993 (acquired) (woven in 1920), collection location unknown (like the Musha style of Datong Village, Ren'ai Township, Nantou County)	Taichung County Cultural Center
NO.61 Shawl for men and women Musha: Renai Township, Nantou County	1934 (collected), Musha, Datong Village, Ren'ai Township, Nantou County	National Taiwan University Museum of Anthropology
NO.62 Woven Mat Musha: Renai Township, Nantou County	1895–1945 (collected), collection location unknown (like the style of Chunyang Village, Ren'ai Township, Nantou County)	Taiwan Museum

(continued)

(continued)

Object name/group affiliation, geographical location	Time and place of museum collection (acquisition)	Sample source
NO.63 Men's sleeveless long coat Pulibau: Xiulin Township and Wanrong Township, Hualien County	2004 (collected), collection location unknown	Taiwan Museum of Prehistory
NO.64 Men's chest pocket Truku Group: Ren'ai Township, Nantou County; Xiulin Township, Wanrong Township, Zhuoxi Township, and Ji'an Township, Hualien County	1993 (collected), Wenlan Village, Xiulin Township, Hualien County	Taiwan Museum of Prehistory
NO.65 Women's Sleeves Truku Group: Ren'ai Township, Nantou County; Xiulin Township, Wanrong Township, Zhuoxi Township, and Ji'an Township, Hualien County	June 1993 (collected) (weaving date 1940), collection location unknown	Taichung County Cultural Center
NO.66 Women's Sleeves Truku Group: Ren'ai Township, Nantou County; Xiulin Township, Wanrong Township, Zhuoxi Township, and Ji'an Township, Hualien County	1992 (collected), Xiulin Village, Xiulin Township, Hualien County	Taiwan Museum of Prehistory
NO.67 Women's Sleeves Truku Group: Ren'ai Township, Nantou County; Xiulin Township, Wanrong Township, Zhuoxi Township, and Ji'an Township, Hualien County	1992 (collected), Xiulin Village, Xiulin Township, Hualien County	Taiwan Museum of Prehistory
NO.68 Women's long skirt Truku Group: Ren'ai Township, Nantou County; Xiulin Township, Wanrong Township, Zhuoxi Township, and Ji'an Township, Hualien County	June 1993 (entered collected) (weaving date 1920), collection location unknown	Taichung County Cultural Center
NO.69 Women's long skirt Truku Group: Ren'ai Township, Nantou County; Xiulin Township, Wanrong Township, Zhuoxi Township, and Ji'an Township, Hualien County	1992 (collected), Shuiyuan Village, Xiulin Township, Hualien County	Taiwan Museum of Prehistory

(continued)

(continued)

Object name/group affiliation, geographical location	Time and place of museum collection (acquisition)	Sample source
NO.70 Shawls for men and women Truku Group: Ren'ai Township, Nantou County; Xiulin Township, Wanrong Township, Zhuoxi Township, and Ji'an Township, Hualien County	June 1998 (acquired by the museum) (weaving date 1937), Wenlan Village, Xiulin Township, Hualien County	Taichung County Cultural Center
NO.71 Shawl for men and women Truku Group: Ren'ai Township, Nantou County; Xiulin Township, Wanrong Township, Zhuoxi Township, and Ji'an Township, Hualien County	June 1998 (acquired by the museum) (weaving year 1940), Xiulin Township, Hualien County	Taichung County Cultural Center
NO.72 Shawl for men and women Truku Group: Ren'ai Township, Nantou County; Xiulin Township, Wanrong Township, Zhuoxi Township, and Ji'an Township, Hualien County	June 1998 (acquired by the museum) (weaving year 1950), Xiulin Township, Hualien County	Taichung County Cultural Center
NO.73 Shawls for men and women Truku Group: Ren'ai Township, Nantou County; Xiulin Township, Wanrong Township, Zhuoxi Township, and Ji'an Township, Hualien County	November 1931 (acquired by the museum), collection location unknown (similar to the style of Xiulin Township, Hualien County)	National Taiwan University Museum of Anthropology
NO.74 Men's sleeveless long coat Unable to identify the subgroup	1895–1945 (in the collection of the museum), collection location unknown (not included in the research sample)	Taiwan Museum
NO.75 Long-sleeved shorts Unable to identify the subgroup	1945 (in the collection of the museum), collection location unknown (not included in the research sample)	Taiwan Museum

Description: Appendix 1 shows 73 research samples obtained for research use from museums and private collectors

Appendix 2
Analysis of the Pattern Types of Atayal Textiles

P. H. Wang and J. Li, *Interpreting the Evolution of Atayal Tribal Textile Patterns*,
https://doi.org/10.1007/978-981-96-5451-2

Group	Sample Number	Object Name	Pattern Type												Subtotal
			Variable Pattern	Diamond Pattern	Circular Pattern	Mortar Pattern	Zigzag Pattern	Square Pattern	Mountain Pattern	Cross Pattern	X Pattern	Net Stripe Pattern	Dot Stripe Pattern	Stripe Pattern	
Hakul (Hakku)	NO.01	Unisex Shawl		●										●	2
Makanji Group	NO.02	Men's Sleeveless Long Robe		●			●		●						3
Msbtunux Group	NO.03	Unisex Shawl		●			●		●					●	4
	NO.04	Unisex Shawl		●	●									●	3
Nan Ao Group	NO.05	Men's Sleeveless Long Shirt		●	●		●	●	●	●	●			●	8
	NO.06	Men's Sleeveless Long Shirt		●				●	●	●				●	5
	NO.07	Men's Sleeveless Long Shirt							●	●	●			●	4
	NO.08	Men's Sleeveless Long Shirt		●					●					●	3
	NO.09	Men's Sleeveless Long Coat							●					●	2
	NO.10	Men's Sleeveless Long Coat		●			●		●					●	4
	NO.11	Men's Sleeveless Long Coat		●							●			●	3

(continued)

(continued)

Group	Sample Number	Object Name	Pattern Type												Subtotal
			Variable Pattern	Diamond Pattern	Circular Pattern	Mortar Pattern	Zigzag Pattern	Square Pattern	Mountain Pattern	Cross Pattern	X Pattern	Net Stripe Pattern	Dot Stripe Pattern	Stripe Pattern	
	NO.12	Men's Sleeveless Long Coat		●						●				●	3
	NO.13	Men's Sleeveless Long Coat		●				●	●	●				●	5
	NO.14	Men's Sleeveless Long Coat		●					●	●	●			●	5
	NO.15	Men's and Women's Long Sleeve Shirts							●					●	3
	NO.16	Men's Long Sleeve Shirts		●				●						●	2
	NO.17	Women's Belts		●			●								2
	NO.18	Women's Belts		●			●		●						3
	NO.19	Women's Belts		●			●								2
	NO.20	Women's Foot Wraps		●						●	●			●	4
	NO.21	Unisex Shawl							●					●	3
	NO.22	Unisex Shawl		●			●		●					●	4
	NO.23	Unisex Shawl		●				●		●				●	4
Marliba Group	NO.24	Men's Sleeveless Long Robe		●					●					●	3

(continued)

(continued)

Group	Sample Number	Object Name	Pattern Type												Subtotal
			Variable Pattern	Diamond Pattern	Circular Pattern	Mortar Pattern	Zigzag Pattern	Square Pattern	Mountain Pattern	Cross Pattern	X Pattern	Net Stripe Pattern	Dot Stripe Pattern	Stripe Pattern	
	NO.25	Men's Chest Wrap		●											1
	NO.26	Tobacco Pouch		●					●						2
Taranan	NO.27	Men's Sleeveless Long Robe										●			1
	NO.28	Women's Long Skirt											●		1
	NO.29	Unisex Shawl			●								●		2
Wan Da Group	NO.30	Men's Long Sleeve Long Robe		●			●		●	●				●	5
Bunsui	NO.31	Men's Sleeveless Long Robe		●			●		●	●				●	5
	NO.32	Men's Sleeveless Long Robe		●						●				●	1
	NO.33	Men's Sleeveless Long Robe		●			●								2
	NO.34	Unisex Long Sleeve Long Robe		●			●		●					●	4
	NO.35	Unisex Long Sleeve Long Robe		●				●	●		●			●	5

(continued)

(continued)

Group	Sample Number	Object Name	Pattern Type												Subtotal
			Variable Pattern	Diamond Pattern	Circular Pattern	Mortar Pattern	Zigzag Pattern	Square Pattern	Mountain Pattern	Cross Pattern	X Pattern	Net Stripe Pattern	Dot Stripe Pattern	Stripe Pattern	
	NO.36	Men's Long Sleeve Long Robe		●			●		●	●	●			●	6
	NO.37	Men's Long Sleeve Short Robe		●										●	2
	NO.38	Men's Chest Wrap		●											1
	NO.39	Women's Long Sleeve Dress		●							●			●	3
	NO.40	Women's Long Sleeve Dress		●	●									●	3
Beishi	NO.41	Men's Long Sleeve Dress		●		●								●	3
	NO.42	Men's Long Sleeve Dress		●										●	2
	NO.43	Men's Open Back Trousers		●					●					●	3
	NO.44	Women's Long Sleeve Dress		●							●			●	3
	NO.45	Women's Long Sleeve Long Dress		●		●					●			●	4
	NO.46	Women's Long Sleeve Short Dress		●										●	2

(continued)

(continued)

Group	Sample Number	Object Name	Pattern Type												Subtotal
			Variable Pattern	Diamond Pattern	Circular Pattern	Mortar Pattern	Zigzag Pattern	Square Pattern	Mountain Pattern	Cross Pattern	X Pattern	Net Stripe Pattern	Dot Stripe Pattern	Stripe Pattern	
	NO.47	Women's Long Sleeve Short Dress		●					●					●	3
	NO.48	Women's Foot Wraps		●	●				●					●	4
	NO.49	Women's Foot Wraps		●	●				●					●	4
	NO.50	Unisex Shawl		●	●				●					●	4
Nan Shi Group	NO.51	Men's Long Sleeve Robe		●					●	●				●	4
	NO.52	Men's Chest Pocket		●											1
	NO.53	Women's Long Skirt	●		●				●	●				●	5
	NO.54	Women's Foot Wraps												●	1
Meicyubus	NO.55	Men's Sleeveless Robe		●			●				●			●	4
	NO.56	Men's Sleeveless Robe		●					●						2
Musha	NO.57	Men's Sleeveless Long Robe		●										●	2
	NO.58	Men's Long Sleeve Long Robe		●			●		●	●	●			●	6

(continued)

(continued)

Group	Sample Number	Object Name	Pattern Type												Subtotal
			Variable Pattern	Diamond Pattern	Circular Pattern	Mortar Pattern	Zigzag Pattern	Square Pattern	Mountain Pattern	Cross Pattern	X Pattern	Net Stripe Pattern	Dot Stripe Pattern	Stripe Pattern	
	NO.59	Women's Long Sleeve Short Robe											●		1
	NO.60	Unisex Shawl		●						●				●	2
	NO.61	Unisex Shawl		●										●	2
	NO.62	Woven Mat		●											1
Pulibau	NO.63	Men's Sleeveless Long Shirt		●		●			●					●	4
Truku Group	NO.64	Men's Chest Wrap		●											1
	NO.65	Women's Sleeve Set		●		●								●	3
	NO.66	Women's Sleeve Set		●					●					●	3
	NO.67	Women's Sleeve Set		●										●	2
	NO.68	Women's Long Skirt		●											1
	NO.69	Women's Long Skirt		●											1
	NO.70	Unisex Shawl		●											1
	NO.71	Unisex Shawl		●											1
	NO.72	Unisex Shawl		●								●			2
	NO.73	Unisex Shawl		●											1
Number of Occurrences			1	63	8	4	15	6	32	16	12	2	3	51	213
Percentage of Occurrences %			0.5	29.6	3.8	1.9	7.0	2.8	15.0	7.5	5.6	0.9	1.4	23.9	100.0

(continued)

(continued)

Group	Sample Number	Object Name	Pattern Type												Subtotal
			Variable Pattern	Diamond Pattern	Circular Pattern	Mortar Pattern	Zigzag Pattern	Square Pattern	Mountain Pattern	Cross Pattern	X Pattern	Net Stripe Pattern	Dot Stripe Pattern	Stripe Pattern	

Description: Appendix 2 shows the pattern types of 73 Atayal textiles. These textile patterns include changing patterns, diamond patterns, circular patterns, mortar-shaped patterns, zigzag patterns, square patterns, mountain patterns, cross patterns, × -shaped patterns, mesh stripes, dot stripes, and stripes, totaling 12 pattern types

Appendix 3
Basic Shape Classification of Atayal Tribal Textile Patterns

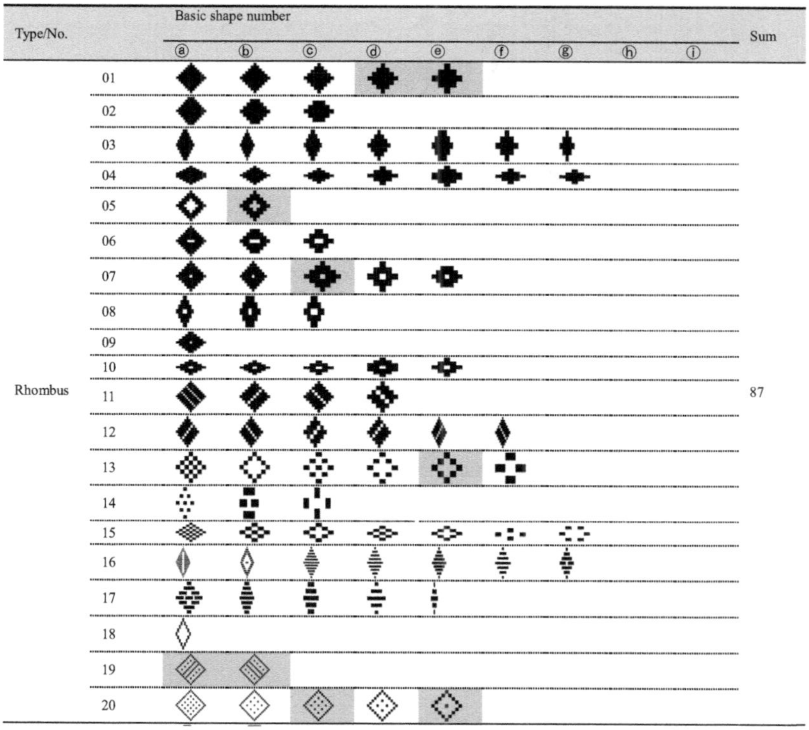

P. H. Wang and J. Li, *Interpreting the Evolution of Atayal Tribal Textile Patterns*, https://doi.org/10.1007/978-981-96-5451-2

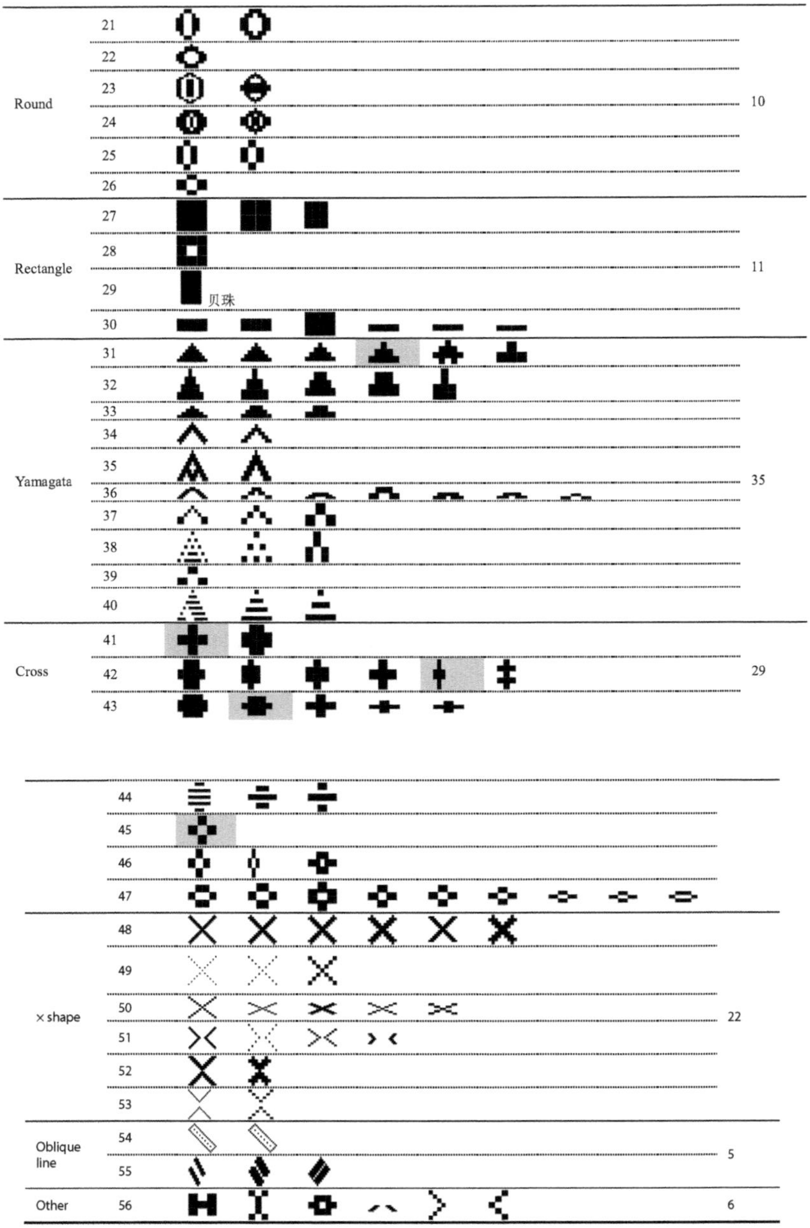

Description: Appendix 3 shows that 774 basic shapes can be extracted from the patterns of the research samples, which can be further divided into 205 basic forms. According to their appearance, they can be summarized into types such as diamond, circle, rectangle, mountain, cross, ×, and oblique line. Diamonds, mountains, and crosses are the most common basic shapes

Appendix 4
Analysis of the Color of the Composite Stripes of Atayal Fabrics

P. H. Wang and J. Li, *Interpreting the Evolution of Atayal Tribal Textile Patterns*,
https://doi.org/10.1007/978-981-96-5451-2

Subgroup	Sample NO	Object Name	Primary Color	Composite stripe color combination												Sum
				No color			Colorful									
				Black	Gray	White	Red	Peach pink	Pink	Orange	Yellow	Green	Blue	Purple	Brown	
Hakul (Hakku)	NO.01 ⓐ	Shawls for men and women	Black and Red				●			●						2
	NO.01 ⓑ						●			●						2
Makanji	NO.02-Front ⓐ	Men's sleeveless long coat	Gray and Peach Pink		●			●								2
	NO.02-Back ⓐ				●			●								2
	NO.02-Back ⓑ				●			●								2
Msbtunux	NO.03 ⓐ	Shawls for men and women	Red	●							●					2
	NO.03 ⓑ			●	●											2
	NO.03 ⓒ			●	●										●	3
	NO.03 ⓓ			●	●											2
	NO.04 ⓐ	Shawls for men and women	Black and Red	●							●					2
	NO.04 ⓑ			●		●										2
	NO.04 ⓒ					●	●									2
Klesan	NO.05-Front ⓐ	Men's sleeveless long coat	Gray and Red		●		●						●			3
	NO.05-Front ⓑ				●		●						●			3
	NO.05-Front ⓒ				●		●						●			3
	NO.05-Front ⓓ				●		●						●			3
	NO.05-Front ⓔ				●								●			2
	NO.05-Front ⓕ				●		●						●			3
	NO.05-Back ⓐ				●		●						●			3
	NO.05-Back ⓑ				●		●						●			3
	NO.05-Back ⓒ				●		●						●			3
	NO.05-Back ⓓ				●		●						●			3
	NO.05- Back ⓔ				●		●						●			3
	NO.05- Back ⓕ				●		●						●			3
	NO.05- Back ⓖ				●		●									2
	NO.05- Back ⓗ				●		●									2
	NO.05-Back ⓘ				●		●									2

(continued)

(continued)

Subgroup	Sample NO	Object Name	Primary Color	Composite stripe color combination												Sum
				No color			Colorful									
				Black	Gray	White	Red	Peach pink	Pink	Orange	Yellow	Green	Blue	Purple	Brown	
	NO.06-Front ⓐ	Men's sleeveless long coat	Gray and Red		●		●						●			3
	NO.06-Front ⓑ				●		●						●			3
	NO.06-Front ⓒ				●		●						●			3
	NO.06-Front ⓓ				●		●						●			3
	NO.06-Front ⓔ				●								●			2
	NO.06-Front ⓕ				●		●						●			3
	NO.06- Back ⓐ				●		●						●			3
	NO.06- Back ⓑ				●								●			2
	NO.06- Back ⓒ				●		●									2
	NO.06- Back ⓓ			●	●											2
	NO.06- Back ⓔ				●		●									2
	NO.07-Front ⓐ	Men's sleeveless long coat	Gray, Red and Blue		●								●			2
	NO.07- Back ⓐ				●		●									2
	NO.07- Back ⓑ				●								●			2
	NO.07- Back ⓒ				●								●			2
	NO.07- Back ⓓ				●		●									2
	NO.08-Front ⓐ	Men's sleeveless long coat	White and Red			●	●									2
	NO.08-Back ⓐ					●	●						●			3
	NO.08- Back ⓑ					●	●									2
	NO.09-Back ⓐ	Men's sleeveless long coat	Gray and Red		●		●									2
	NO.10-Front ⓐ	Men's sleeveless long coat	Black and Peach Pink		●			●				●	●			4
	NO.10-Back ⓐ				●			●				●	●			4
	NO.10-Back ⓑ				●			●				●	●			4
	NO.10-Back ⓒ				●			●				●	●			4
	NO.10-Back ⓓ			●	●											2

(continued)

(continued)

Subgroup	Sample NO	Object Name	Primary Color	Composite stripe color combination												Sum
				No color			Colorful									
				Black	Gray	White	Red	Peach pink	Pink	Orange	Yellow	Green	Blue	Purple	Brown	
	NO.10-Backⓔ			●	●											2
	NO.10-Backⓕ			●	●											2
	NO.10-Backⓖ			●	●											2
	NO.10-Backⓗ			●	●											2
	NO.10-Backⓘ			●	●											2
	NO.11-Frontⓐ	Men's sleeveless long coat	Black and Red	●		●	●									3
	NO.11-Frontⓑ			●			●									2
	NO.11-Backⓐ			●		●	●									3
	NO.11-Backⓑ			●	●											2
	NO.11-Backⓒ			●	●											2
	NO.11-Backⓓ			●	●											2
	NO.11-Backⓔ			●	●											2
	NO.11-Backⓕ			●	●											2
	NO.11-Backⓖ			●	●											2
	NO.11-Backⓗ			●	●											2
	NO.11-Backⓘ			●	●											2
	NO.11-Backⓙ			●	●											2
	NO.11-Backⓚ			●	●											2
	NO.11-Backⓛ			●	●											2
	NO.12-Frontⓐ	Men's sleeveless long coat	Black, Gray and Red		●		●									2
	NO.12-Backⓐ			●	●		●									3
	NO.12-Backⓑ			●	●											2
	NO.12-Backⓒ			●	●											2
	NO.12-Backⓓ			●	●											2
	NO.12-Backⓔ			●	●											2
	NO.12-Backⓕ			●	●											2

(continued)

(continued)

Subgroup	Sample NO	Object Name	Primary Color	Composite stripe color combination												Sum
				No color			Colorful									
				Black	Gray	White	Red	Peach pink	Pink	Orange	Yellow	Green	Blue	Purple	Brown	
	NO.12-Backⓖ			●	●											2
	NO.12-Backⓗ			●	●											2
	NO.13-Frontⓐ	Men's sleeveless shorts	Gray and Red		●		●						●			3
	NO.13-Frontⓑ				●		●						●			3
	NO.13-Backⓐ					●	●						●			3
	NO.13-Backⓑ					●	●						●			3
	NO.14-Frontⓐ	Men's sleeveless shorts	White, Red			●							●			2
	NO.14-Frontⓑ					●	●						●			3
	NO.14-Frontⓒ					●	●						●			3
	NO.14-Backⓐ					●	●						●			3
	NO.14-Backⓑ					●	●									2
	NO.15-Frontⓐ	Men's and women's long-sleeved shorts	Black and Red	●			●						●			3
	NO.16-Frontⓐ	Men's long-sleeved shirt	Red and Blue		●		●									2
	NO.16-Frontⓑ				●								●			2
	NO.16-Frontⓒ				●		●						●			3
	NO.16-Frontⓓ				●		●						●			3
	NO.16-Frontⓔ				●								●			2
	NO.16-Frontⓕ				●		●						●			3
	NO.16-Frontⓖ				●								●			2
	NO.17ⓐ	Women's belts	Black, Red and Blue	●	●		●									3
	NO.17ⓑ			●	●								●			3
	NO.17ⓒ			●	●		●									3
	NO.17ⓓ			●	●								●			3
	NO.17ⓔ			●	●		●									3
	NO.17ⓕ			●	●		●									3

(continued)

(continued)

Subgroup	Sample NO	Object Name	Primary Color	Composite stripe color combination												Sum
				No color			Colorful									
				Black	Gray	White	Red	Peach pink	Pink	Orange	Yellow	Green	Blue	Purple	Brown	
	NO.17ⓖ			●	●		●									3
	NO.17ⓗ			●	●		●									3
	NO.17ⓘ			●	●								●			3
	NO.17ⓙ			●	●		●									3
	NO.17ⓚ			●	●		●									3
	NO.17ⓛ			●	●		●									3
	NO.18ⓐ	Women's belts	Black	●	●										●	3
	NO.18ⓑ			●	●										●	3
	NO.18ⓒ			●	●										●	3
	NO.18ⓓ			●	●										●	3
	NO.18ⓔ			●	●										●	3
	NO.18ⓕ			●	●										●	3
	NO.18ⓖ			●	●										●	3
	NO.18ⓗ			●	●										●	3
	NO.18ⓘ			●	●										●	3
	NO.18ⓙ			●	●										●	3
	NO.18ⓚ			●	●										●	3
	NO.18ⓛ			●	●										●	3
	NO.18ⓜ			●	●										●	3
	NO.19ⓐ	Women's belts	Yellow and Blue			●					●	●	●			4
	NO.20ⓐ	Women's foot protection cloth	Yellow and Blue		●								●			2
	NO.20ⓑ				●						●		●			3
	NO.20ⓒ				●						●		●			3
	NO.21ⓐ	Shawls for men and women	Black,Peach Pink, and Yellow	●				●					●			3
	NO.21ⓑ			●				●			●	●				4
	NO.22ⓐ	Shawls for men and women	Black and Peach Pink	●				●			●		●			4

(continued)

(continued)

Subgroup	Sample NO	Object Name	Primary Color	Composite stripe color combination												Sum
				No color			Colorful									
				Black	Gray	White	Red	Peach pink	Pink	Orange	Yellow	Green	Blue	Purple	Brown	
	NO.22ⓑ			●				●					●	●		4
	NO.22ⓒ			●				●					●	●		4
	NO.22ⓓ			●				●					●	●		4
	NO.23ⓐ	Shawls for men and women	Red and Blue			●							●			2
	NO.23ⓑ					●	●									2
Malepa	NO.24-Backⓐ	Men's sleeveless long coat	Gray, Red and Blue			●	●						●			3
	NO.24-Backⓑ					●	●									2
	NO.25ⓐ	Men's chest pocket	Red and Blue		●		●						●			3
	NO.26ⓐ	Tobacco Pouch	Black, Gray and Red	●	●											2
	NO.26ⓑ			●	●		●									3
	NO.26ⓒ					●	●									2
	NO.26ⓓ			●	●											2
Curved group	NO.27-Frontⓐ	Men's sleeveless long coat	Gray and Blue		●								●			2
	NO.27-Frontⓑ				●								●			2
	NO.28ⓐ	Women's long skirt	Black	●		●										2
	NO.29ⓐ	Shawls for men and women	Black	●		●										2
Plnawan	NO.30-Frontⓐ	Men's long sleeve coat	Red			●	●						●			3
	NO.30-Frontⓑ					●							●			2
	NO.30-Backⓐ					●							●			2
	NO.30-Backⓑ			●		●										2
	NO.30-Backⓒ						●						●			2
Bunsui	NO.31-Backⓐ	Men's long sleeve coat	Gray and Red		●		●									2
	NO.31-Backⓑ				●		●						●			3
	NO.31-Backⓒ				●								●			2
	NO.31-Backⓓ				●		●						●			3

(continued)

(continued)

Subgroup	Sample NO	Object Name	Primary Color	Composite stripe color combination												Sum
				No color		Colorful										
				Black	Gray	White	Red	Peach pink	Pink	Orange	Yellow	Green	Blue	Purple	Brown	
	NO.31-Backⓔ				●		●						●			3
	NO.31-Backⓕ				●		●						●			3
	NO.32-Frontⓐ	Men's long sleeve coat	Gray and Red		●		●									2
	NO.32-Frontⓑ			●	●											2
	NO.32-Frontⓒ			●	●		●									3
	NO.32-Frontⓓ				●		●						●			3
	NO.32-Backⓐ			●	●		●									3
	NO.32-Backⓑ				●		●						●			3
	NO.33-Backⓐ	Men's long sleeve coat	Red			●	●									2
	NO.33-Backⓑ					●	●									2
	NO.34-Frontⓐ	Men's and women's long sleeves	Black, Gray and Red		●		●					●	●			4
	NO.34-Frontⓑ				●		●					●	●			4
	NO.34-Frontⓒ				●		●									2
	NO.34-Frontⓓ				●		●						●			3
	NO.34-Frontⓔ				●								●			2
	NO.34-Backⓐ				●		●						●			3
	NO.34-Backⓑ				●		●						●			3
	NO.35-Frontⓐ	Men's long sleeve coat	Black, White and Red			●							●			2
	NO.35-Frontⓑ					●	●									2
	NO.35-Frontⓒ					●	●		●				●			4
	NO.35-Frontⓓ					●	●		●				●			4
	NO.35-Backⓐ					●	●						●			3
	NO.35-Backⓑ					●	●		●				●			4
	NO.35-Backⓒ					●	●		●				●			4
	NO.36-Frontⓐ	Men's long sleeve coat	Black, Gray and Blue			●							●			2
	NO.36-Frontⓑ					●			●		●		●			4

(continued)

(continued)

Subgroup	Sample NO	Object Name	Primary Color	Composite stripe color combination												Sum
				No color		Colorful										
				Black	Gray	White	Red	Peach pink	Pink	Orange	Yellow	Green	Blue	Purple	Brown	
	NO.36-Frontⓒ					●	●					●				3
	NO.36-Frontⓓ					●	●				●	●				4
	NO.36-Frontⓔ					●	●					●				3
	NO.36-Backⓐ					●	●				●	●	●			5
	NO.36-Backⓑ					●	●						●			3
	NO.37-Frontⓐ	Men's long-sleeved shirt	Black, Gray and Red	●	●		●									3
	NO.37-Frontⓑ			●	●								●			3
	NO.37-Frontⓒ			●	●											2
	NO.37-Frontⓓ			●	●											2
	NO.37-Frontⓔ			●	●											2
	NO.37-Backⓐ			●	●		●									3
	NO.37-Backⓑ				●		●						●			3
	NO.38ⓐ	Men's chest pocket	Black and Orange			●		●		●			●			4
	NO.39-Frontⓐ	Women's long sleeve dress	Red and Orange	●		●	●		●	●	●				●	7
	NO.39-Frontⓑ					●				●						2
	NO.39-Frontⓒ			●					●		●					3
	NO.39-Frontⓓ			●		●				●						3
	NO.39-Frontⓔ					●			●	●						3
	NO.39-Frontⓕ				●	●			●	●	●		●		●	7
	NO.39-Backⓐ				●	●	●		●		●					5
	NO.39-Backⓑ			●		●	●		●	●	●					6
	NO.39-Backⓒ					●	●		●	●	●					5
	NO.39-Backⓓ			●		●	●		●	●	●					6
	NO.40-Frontⓐ	Women's long sleeve dress	Red			●	●		●		●		●		●	6
	NO.40-Frontⓑ					●	●									2
	NO.40-Frontⓒ								●	●	●		●		●	5

(continued)

(continued)

Subgroup	Sample NO	Object Name	Primary Color	Composite stripe color combination												Sum
				No color			Colorful									
				Black	Gray	White	Red	Peach pink	Pink	Orange	Yellow	Green	Blue	Purple	Brown	
	NO.40-Front ⓓ					●	●					●	●		●	5
	NO.40-Front ⓔ			●		●	●		●		●	●		●		7
	NO.40-Back ⓐ					●	●		●	●		●	●			6
	NO.40-Back ⓑ			●		●	●		●	●		●				6
Beishi	NO.41-Front ⓐ	Men's long sleeve coat	Black, Gray and Red	●		●	●			●		●	●			6
	NO.41-Front ⓑ			●		●	●					●	●			5
	NO.42-Front ⓐ	Men's long sleeve coat	Black, Gray and Red	●	●	●	●		●	●			●			7
	NO.42-Front ⓑ				●		●									2
	NO.43 ⓐ	Men's open back pants	Black, Gray and Red	●	●			●	●		●	●	●			7
	NO.44-Front ⓐ	Women's long sleeve dress	Red	●		●	●									3
	NO.44-Front ⓑ					●	●									2
	NO.44-Front ⓒ					●	●									2
	NO.44-Front ⓓ					●	●									2
	NO.44-Front ⓔ					●	●									2
	NO.44-Front ⓕ			●		●	●	●	●		●		●			7
	NO.44-Front ⓖ					●	●									2
	NO.44-Front ⓗ			●		●										2
	NO.44-Front ⓘ					●			●				●			3
	NO.44-Front ⓙ			●		●	●									3
	NO.44-Back ⓐ			●		●	●		●			●				5
	NO.44-Back ⓑ			●		●	●									3
	NO.44-Back ⓒ			●		●										2
	NO.44-Back ⓓ			●		●	●									3
	NO.44-Back ⓔ			●		●	●									3
	NO.44-Back ⓕ					●	●									2

(continued)

(continued)

Subgroup	Sample NO	Object Name	Primary Color	Composite stripe color combination												Sum
				No color			Colorful									
				Black	Gray	White	Red	Peach pink	Pink	Orange	Yellow	Green	Blue	Purple	Brown	
	NO.44-Back ⓖ			●		●							●			3
	NO.45-Front ⓐ	Women's long-sleeved shirt	Black and Red	●		●	●									3
	NO.45-Front ⓑ					●	●									2
	NO.45-Front ⓒ			●		●	●	●								4
	NO.45-Front ⓓ			●		●	●									3
	NO.46-Front ⓐ	Women's long-sleeved shirt	Black, Gray and Red	●	●		●		●		●	●				6
	NO.46-Front ⓑ			●	●		●		●	●	●		●			7
	NO.46-Front ⓒ			●	●		●		●		●	●	●			7
	NO.47-Front ⓐ	Women's long-sleeved shirt	Black, White and Red			●	●				●			●	●	5
	NO.47-Front ⓑ									●	●			●	●	4
	NO.47-Front ⓒ			●		●	●			●	●			●		6
	NO.47-Front ⓓ			●		●	●			●	●			●		6
	NO.48 ⓐ	Women's foot protection cloth	Gray and Blue	●							●				●	3
	NO.48 ⓑ			●		●						●	●	●		5
	NO.49 ⓐ	Women's foot protection cloth	Black, Gray and Yellow	●	●						●				●	4
	NO.49 ⓑ			●								●	●			3
	NO.50 ⓐ	Shawls for men and women	Yellow and Blue	●						●		●	●			4
	NO.50 ⓑ			●		●	●				●					4
	NO.50 ⓒ			●		●						●				3
	NO.50 ⓓ			●			●	●		●	●					5
Nanshi	NO.51-Front ⓐ	Men's long sleeve coat	Gray and Red		●		●									2
	NO.51-Front ⓑ				●		●						●			3
	NO.51-Front ⓒ				●		●				●	●	●			5
	NO.51-Back ⓐ				●		●									2
	NO.51-Back ⓑ				●		●									2
	NO.51-Back ⓒ				●		●									2

(continued)

(continued)

Subgroup	Sample NO	Object Name	Primary Color	Composite stripe color combination												Sum
				No color			Colorful									
				Black	Gray	White	Red	Peach pink	Pink	Orange	Yellow	Green	Blue	Purple	Brown	
	NO.52ⓐ	Men's chest pocket	Red			●	●									2
	NO.53ⓐ	Women's long skirt	Black and Red	●									●			2
	NO.53ⓑ			●				●		●						3
	NO.53ⓒ			●									●			2
	NO.53ⓓ			●				●		●						3
	NO.54ⓐ	Women's foot protection cloth	White			●	●						●			3
Meicyubus	NO.55-Front ⓐ	Men's sleeveless long coat	Gray and Red		●		●						●			3
	NO.55-Frontⓑ				●		●						●			3
	NO.55-Frontⓒ				●		●						●			3
	NO.55-Backⓐ				●		●						●			3
	NO.56-Frontⓐ	Men's sleeveless long coat	Gray and Red		●								●			2
	NO.56-Frontⓑ				●		●						●			3
	NO.56-Backⓐ				●		●									2
	NO.56-Backⓑ				●		●									2
Musha	NO.57-Frontⓐ	Men's sleeveless long coat	Red	●	●		●									3
	NO.57-Frontⓑ			●		●	●									3
	NO.58-Frontⓐ	Men's long sleeve coat	Black and Red	●		●										2
	NO.58-Frontⓑ			●		●										2
	NO.58-Frontⓒ			●		●										2
	NO.58-Frontⓓ			●		●										2
	NO.58-Frontⓔ			●		●										2
	NO.58-Frontⓕ			●		●										2
	NO.58-Frontⓖ			●		●										2
	NO.58-Frontⓗ			●		●										2
	NO.58-Frontⓘ			●		●										2

(continued)

(continued)

Subgroup	Sample NO	Object Name	Primary Color	Composite stripe color combination												Sum
				No color			Colorful									
				Black	Gray	White	Red	Peach pink	Pink	Orange	Yellow	Green	Blue	Purple	Brown	
	NO.58-Frontⓙ			●		●										2
	NO.58-Frontⓚ			●		●										2
	NO.58-Frontⓛ			●		●										2
	NO.58-Frontⓜ			●		●										2
	NO.58-Frontⓝ			●		●										2
	NO.58-Frontⓞ			●		●										2
	NO.58-Frontⓟ			●		●										2
	NO.58-Frontⓠ			●		●										2
	NO.58-Frontⓡ			●		●										2
	NO.58-Frontⓢ			●		●										2
	NO.58-Frontⓣ			●		●										2
	NO.58-Frontⓤ			●		●										2
	NO.58-Frontⓥ			●		●										2
	NO.58-Frontⓦ			●		●										2
	NO.58-Frontⓧ			●		●										2
	NO.58-Frontⓦ			●		●										2
	NO.58-Frontⓧ			●		●										2
	NO.58-Frontⓨ			●		●										2
	NO.58-Backⓐ			●		●										2
	NO.58-Backⓑ			●		●										2
	NO.58-Backⓒ			●		●										2
	NO.58-Backⓓ			●		●										2
	NO.58-Backⓔ			●		●										2
	NO.58-Backⓕ			●		●										2
	NO.58-Backⓖ			●		●										2
	NO.58-Backⓗ			●		●										2

(continued)

(continued)

Subgroup	Sample NO	Object Name	Primary Color	Composite stripe color combination												Sum
				No color			Colorful									
				Black	Gray	White	Red	Peach pink	Pink	Orange	Yellow	Green	Blue	Purple	Brown	
	NO.58-Backⓘ			●		●										2
	NO.58-Backⓙ			●		●										2
	NO.58-Backⓚ			●		●										2
	NO.58-Backⓛ			●		●										2
	NO.58-Backⓜ			●		●										2
	NO.58-Backⓝ			●		●										2
	NO.58-Backⓞ			●		●										2
	NO.58-Backⓟ			●		●										2
	NO.58-Backⓠ			●		●										2
	NO.58-Backⓡ			●		●										2
	NO.58-Backⓢ			●		●										2
	NO.59-Frontⓐ	Women's long-sleeved shirt	Black and Red	●		●	●									3
	NO.59-Frontⓑ			●		●	●						●			4
	NO.60ⓐ	Shawls for men and women	Black and Red	●			●					●				3
	NO.60ⓑ						●			●						2
	NO.60ⓒ						●			●						2
	NO.60ⓓ						●			●						2
	NO.61ⓐ1–4	Shawls for men and women	Black and Red	●			●									2
	NO.61ⓑ1–4			●			●									2
	NO.61ⓒ1–5			●			●									2
	NO.61ⓓ1–5			●			●									2
	NO.61ⓔ1–5			●			●									2
	NO.61ⓕ1–5			●			●									2
	NO.61ⓖ1–5			●			●									2
	NO.61ⓗ1–5			●			●									2
	NO.62ⓐ	Weaving Mat	Brown	●	●										●	3

(continued)

(continued)

Subgroup	Sample NO	Object Name	Primary Color	Composite stripe color combination												Sum
				No color			Colorful									
				Black	Gray	White	Red	Peach pink	Pink	Orange	Yellow	Green	Blue	Purple	Brown	
	NO.62ⓑ				●										●	2
	NO.62ⓒ				●										●	2
Pulibau	NO.63-Frontⓐ	Men's sleeveless long coat	Black, Gray and Red	●			●									2
	NO.63-Frontⓑ			●		●										2
	NO.63-Frontⓒ					●	●									2
	NO.63-Frontⓓ					●	●									2
	NO.63-Frontⓔ					●	●									2
	NO.63-Backⓐ			●		●										2
	NO.63-Backⓑ			●		●										2
	NO.63-Backⓒ			●		●										2
	NO.63-Backⓓ			●		●										2
	NO.63-Backⓔ			●		●										2
	NO.63-Backⓕ			●		●										2
	NO.63-Backⓖ			●		●										2
	NO.63-Backⓗ			●		●										2
Truku Group	NO.64ⓐ	Men's chest pocket	Red			●	●									2
	NO.65ⓐ	Women's Suits	White			●	●					●				3
	NO.65ⓑ			●		●	●					●				4
	NO.66ⓐ	Women's Suits	White			●	●									2
	NO.66ⓑ					●							●			2
	NO.67ⓐ	Women's Suits	Gray	●	●			●								3
	NO.67ⓑ			●	●			●								3
	NO.68ⓐ	Women's long skirt	White			●						●		●		3
	NO.68ⓑ					●						●		●		3
	NO.68ⓒ					●						●		●		3
	NO.69ⓐ	Women's long skirt	White and Red			●	●						●			3

(continued)

(continued)

Subgroup	Sample NO	Object Name	Primary Color	Composite stripe color combination												Sum
				No color			Colorful									
				Black	Gray	White	Red	Peach pink	Pink	Orange	Yellow	Green	Blue	Purple	Brown	
	NO.69ⓑ					●	●					●	●			4
	NO.70ⓐ	Shawls for men and women	White, Red and Green		●		●					●	●			4
	NO.71ⓐ	Shawls for men and women	White, Red and Blue		●		●					●	●	●		5
	NO.71ⓑ				●		●					●	●	●		5
	NO.71ⓒ				●		●	●				●	●	●		6
	NO.72ⓐ	Shawls for men and women	Black, Gray and Red	●	●		●									3
	NO.72ⓑ				●		●	●								3
	NO.72ⓒ			●	●		●									3
	NO.73ⓐ	Shawls for men and women	White and Red Blue			●	●						●			3
Occurrence				196	173	160	187	24	26	27	34	38	127	15	26	104
Ratio (%)				19.0	16.7	15.5	18.1	2.3	2.5	2.6	3.3	3.7	12.3	1.5	2.5	100

Description: Appendix 4 shows the color combination analysis of the composite stripes of 73 Atayal fabrics. It is found that the achromatic colors of black, gray, and white, as well as the colorful shades of red, pink, pink, orange, yellow, green, blue, purple, and brown, are all standard colors used in Atayal fabrics

Postscript

Regarding the narrative of the research conclusions, although there are phased results, there is still a need for reflection and expectation. It is hoped that through this research process and results, we can understand Taiwan's ethnic minorities and pay attention to their material culture research. It is also expected to connect the vision of design studies with cultural research and then continue the future development of Taiwan's humanities and arts. The following are thoughts on subsequent research from teaching, practice, and research.

For a long time, design education has mainly inherited Western art and design thinking, especially the content of introductory formative design courses, usually taking Bauhaus design education as a reference. However, in recent years, with the enthusiasm of Eastern culture and primitive art, there has also been a discussion of ethnic style. Although Eastern culture or primitive art does not define the term "design," the appearance and decorative beauty presented by its material culture, in addition to symbolizing the unique cultural image of the nation, seems to indicate that the behavior of design has long existed. It is expected that the Atayal tribe's textile patterns analyzed in this study can be discussed in the formative design course in the future.

The current society has realized the importance of "traditional wisdom creation of ethnic minorities," and relevant units also protect the traditional culture of Taiwan's ethnic minorities through legislation. Relatively, it may also lead to restrictions on the creation of artists who are not ethnic minorities due to legislative protection, and it is currently difficult to clarify what the traditional art of Taiwan's ethnic minorities is. How can we separate ourselves from the artistic expression of ethnic style? Therefore, in addition to more actively and specifically organizing the traditional art of Taiwan's ethnic minorities, adopting a more open view, and encouraging the application of elements of ethnic minority traditional culture so that the cultural art of Taiwan's ethnic minorities can be internationalized and become a representative cultural industry of Taiwan's region.

P. H. Wang and J. Li, *Interpreting the Evolution of Atayal Tribal Textile Patterns*,
https://doi.org/10.1007/978-981-96-5451-2

In the Atayal tribe, men and women have different daily tasks. Men are responsible for hunting, fighting, and farming, while women are responsible for weaving, cooking, child-rearing, sowing, and harvesting. Weaving is the duty of Atayal women, and the traditional horizontal weaving machine is made by the women's fathers and elders or commissioned by skilled makers. It is worth noting that men are not supposed to be familiar with weaving techniques, but men make the tools. It is speculated that these items are produced jointly by men and women after communication and coordination. The Atayal tribe uses traditional clothing as part of its cultural heritage. In cultural transmission, the objects represented by the woven patterns have transcended physical reality, mainly connected to ancestral beliefs, spiritual life, and social power, and continuously changed. Geometric woven patterns cannot exist independently but are attached to clothing. In the primitive cultural stage, geometric patterns have important life significance, derived from the tribe's understanding of natural phenomena, and then obtained primary refinement to form symbolic characteristics. The Atayal tribe's woven patterns result from women's careful design; determining the pattern style is the shaping, basic form, and the composition of compound stripes; women dominate the overall artistic form of expression. Compared to other Taiwanese minorities, the Paiwan tribe's artistic creation is more balanced, with men specializing in wood carving and house building; women are good at embroidery and sewing, their tribal environment displays aesthetic experience and artistic potential everywhere, the overall division of labor seems complete, and the tribal style is so harmonious. For future research ideas, perhaps the creative expressions of the Atayal and Paiwan tribes can be compared, and the differences in interpreting patterns among the tribes can be explored through comparative research, or even how the division of labor between the sexes affects artistic development from a gender perspective. The Atayal tribe's woven patterns are an essential part of Taiwan's minority material culture, showing the Atayal tribe's aesthetic concepts and lifestyle and displaying unique abstract geometric aesthetics. Although this study has analyzed the Atayal tribe's woven patterns, trying to dig deeper into existing woven patterns to find discoveries and propose different research results from the past, this is just one of many minorities, and there are still topics worth paying attention to, such as the material culture of the Austronesian people.

Moreover, there are many unique minorities within our country, and their textiles also have decorative solid properties, such as the traditional costumes of the Miao and Dong tribes. These minorities with unique cultural backgrounds should also be valued, especially in the face of the development of modern society, where traditional, unique cultures may gradually wither. If it can be recorded through research methods, it will undoubtedly add a new chapter to "Design Humanities and Art."

Wang Po Hsun

September 2023 in Cotai, Macau

GPSR Compliance

The European Union's (EU) General Product Safety Regulation (GPSR) is a set of rules that requires consumer products to be safe and our obligations to ensure this.

If you have any concerns about our products, you can contact us on ProductSafety@springernature.com

In case Publisher is established outside the EU, the EU authorized representative is:

Springer Nature Customer Service Center GmbH
Europaplatz 3
69115 Heidelberg, Germany

Batch number: 10704307

Printed by Printforce, the Netherlands